KT-406-468

Collins

NEW GCSE MATHS
Edexcel Linear
Fully supports the 2010 GCSE Specification

Brian Speed • Keith Gordon • Kevin Evans
Trevor Senior • Chris Pearce

CONTENTS

INTRODUCTION

Welcome to Collins New GCSE Maths for Edexcel Linear Higher Homework Book 1. This book follows the structure of the Edexcel Linear Higher Student Book 1.

Colour-coded grades

Know what target grade you are working at and track your progress with the colour-coded grade panels at the side of the page.

Use of calculators

Questions when you could use a calculator are marked with a ✓ icon.

Examples

Recap on methods you need by reading through the examples before starting the homework exercises.

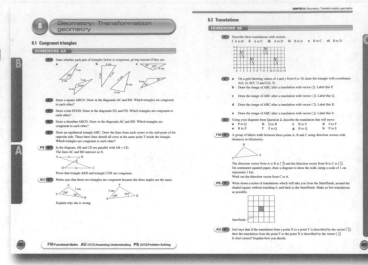

Functional maths

Practise functional maths skills to see how people use maths in everyday life. Look out for practice questions marked **FM**.

There are also extra functional maths and problem-solving activities at the end of every chapter to build and apply your skills.

New Assessment Objectives

Practise new parts of the curriculum (Assessment Objectives AO2 and AO3) with questions that assess your understanding

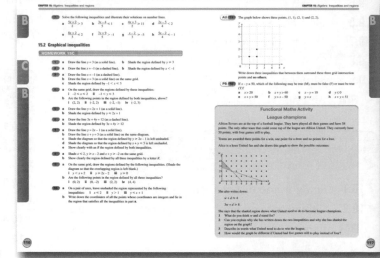

marked **AU** and questions that test if you can solve problems marked **PS**. You will also practise some questions that involve several steps and where you have to choose which method to use; these also test AO2. There are also plenty of straightforward questions (AO1) that test if you can do the maths.

Student Book CD-ROM

Remind yourself of the work covered in class with the Student Book in electronic form on the CD-ROM. Insert the CD into your machine and choose the chapter you need.

1 Number: Number skills and properties

1.1 Solving real-life problems

HOMEWORK 1A

FM 1 Andy needs enough tiles to cover 12 m² in his bathroom. It takes 25 tiles to cover 1 m².
It is recommended to buy 20 percent more tiles to allow for cutting.
Tiles are sold in boxes of 16. Andy buys 24 boxes. Does he have enough tiles?

FM 2 The organiser of a church fête needs 1000 balloons. She has a budget of £30.
Each packet contains 25 balloons and costs 85p. Does she have enough?

FM 3 A TV rental shop buys TVs for £110 each.
The shop needs to make a least 10 percent profit on each TV to cover its costs.
On average each TV is rented for 40 weeks at £3.50 per week.
Does the shop cover its costs?

4 The annual subscription fee to join a fishing club is £42. The treasurer of the club has
collected £1134 in fees. How many people have paid their subscription fee?

5 Mrs Woodhead saves £14 per week towards her bills. How much does she save in a year?

FM 6 Mark saves £15 each week.
He wants to buy a dining set costing £860.
Will he have saved enough money after one year?
Show how you worked out your answer.

7 Sylvia has a part-time job and is paid £18 for every day she works. Last year she worked
for 148 days. How much was she paid for the year?

PS 8 Mutya has a part-time job, working three days each week.
She is paid £5 per hour.
She works 4 hours each day.
Neil has a full-time job, working five days each week.
He is paid £6 per hour.
He works 7 hours each day.
How many weeks does Mutya have to work to earn at least as much as Neil earns in one
week?
Show how you worked out your answer.

FM 9 A coach firm charges £504 for 36 people to go Christmas shopping on a day trip to
Calais. The cost of the coach is shared equally between the passengers. Mary takes £150.
When she gets to Calais, Mary wants to buy games costing €50 each for each of her four
grandchildren. The exchange rate is £1 = €1.25. Does she have enough money to pay for
the presents?

10 A concert hall has 48 rows of seats with 32 seats in a row. What is the maximum capacity
of the hall?

AU 11 Allan is a market gardener and has 420 bulbs to plant. He plants them out in rows with
18 bulbs to a row. How many complete rows will there be?

FM Functional Maths   **AU** (AO2) Assessing Understanding   **PS** (AO3) Problem Solving

FM 12 A room measuring 6 m by 8 m is to be carpeted. The carpet costs £19 per m^2.

    a Estimate the cost of the carpet.

    b Calculate the exact cost of the carpet.

AU 13 Paul's room measures 7 m by 3 m.

    The carpet is sold from rolls which are 4 m wide.

    This means that a person buying carpet has to buy pieces from the full width of the roll.

    a What is the smallest area of carpet that Paul has to buy?

    b He has £300 to spend on the carpet.

       What is the most he can spend per m^2?

FM 14 There are 240 students and teachers on a school visit.

    Four coaches are booked.

    Each coach can take 53 passengers.

    The school decides to book another coach for the remaining passengers.

    What is the smallest number of seats needed on this coach?

PS 15 On average, a toy shop sells 38 computer games each week.

    The manager has a delivery of 150 games each month.

    Will his stock of games increase or decrease?

    Show clearly how you decide.

1.2 Multiplication and division with decimals

HOMEWORK 1B

1 Evaluate each of these.

    a 0.5×0.5 **b** 12.6×0.6 **c** 7.2×0.7

    d 1.4×1.2 **e** 2.6×1.5

2 For each of the following:

    i estimate the answer by first rounding off each number to the nearest whole number.

    ii calculate the exact answer, and then, by doing a subtraction, calculate how much out your answer to part **i** is.

    a 3.7×2.4 **b** 4.8×3.1 **c** 5.1×4.2

    d 6.5×2.5

PS AU 3 **a** Use any method to work out 15×16

       Use your answer to work out:

    b **i** 1.5×1.6

       ii 0.75×3.2

       iii 4.5×1.6

PS 4 **a** Work out 7.2×3.4

    b Explain how you can use your answer to part **a** to write down the answer to 6.2×3.4

AU 5 **a** Work out 2.3×7.5

    b Given that $4.1 \times 7.5 = 30.75$, use this fact with your answer to part **a** to work out 6.4×7.5

6 Evaluate each of these.

    a 3.12×14 **b** 5.24×15 **c** 1.36×22

    d 7.53×25 **e** 27.1×32

7 Find the total cost of each of the following purchases.
    a Twenty-four litres of petrol at £0.92 per litre.
    b Eighteen pints of milk at £0.32 per pint.
    c Fourteen magazines at £2.25 per copy.

8 A CD case is 0.8 cm thick.
How many cases are in a pile of CDs that is 16 cm high?

PS 9   **a** Use any method to work out $64 \div 4$
    Use your answer to work out:
    b **i** $6.4 \div 0.04$ **ii** $0.64 \div 4$ **iii** $0.064 \div 0.4$

PS 10 Here are three calculations.
    $43.68 \div 5.6$ $21.7 \div 6.2$ $19.74 \div 2.1$
Which has the largest answer?
Show how you know.

1.3 Approximation of calculations

HOMEWORK 1C

1 Round each of the following to one significant figure.
a 51 203	**b** 56 189	**c** 33 261	**d** 89 998	**e** 94 999
f 53.71	**g** 87.24	**h** 31.06	**i** 97.835	**j** 184.23
k 0.5124	**l** 0.2765	**m** 0.006 12	**n** 0.049 21	**o** 0.000 888
p 9.7	**q** 85.1	**r** 91.86	**s** 196	**t** 987.65

AU 2 What is the least and the greatest number of people that can be found in these towns?
    Hellaby   population 900 (to 1 significant figure)
    Hook   population 650 (to 2 significant figures)
    Hundleton   population 1050 (to 3 significant figures)

3 Round off each of the following numbers to 2 significant figures.
a 6725	**b** 35 724	**c** 68 522	**d** 41 689	**e** 27 308
f 6973	**g** 2174	**h** 958	**i** 439	**j** 327.6

4 Round off each of the following to the number of significant figures (sf) indicated.
a 46 302 (1 sf)	**b** 6177 (2 sf)	**c** 89.67 (3 sf)	**d** 216.7 (2 sf)
e 7.78 (1 sf)	**f** 1.087 (2 sf)	**g** 729.9 (3 sf)	**h** 5821 (1 sf)
i 66.51 (2 sf)	**j** 5.986 (1 sf)	**k** 7.552 (1 sf)	**l** 9.7454 (3 sf)
m 25.76 (2 sf)	**n** 28.53 (1 sf)	**o** 869.89 (3 sf)	**p** 35.88 (1 sf)
q 0.084 71 (2 sf)	**r** 0.0099 (2 sf)	**s** 0.0809 (1 sf)	**t** 0.061 97 (3 sf)

PS 5 A baker estimates that she has baked 100 loaves. She is correct to one significant figure.
She sells two loaves and now has 90 loaves to one significant figure.
How many could she have had to start with?
Work out all possible answers.

AU 6 There are 500 cars in a car park to one significant figure.
What is the least possible number of cars that could enter the car park so that there are 700 cars in the car park to one significant figure?

FM 7 A supermarket manager estimates that five minutes before closing time there are 200 people still shopping.

This number is to one significant figure.

On average a checkout serves four customers in five minutes.

No more shoppers enter the supermarket.

How many checkouts should he open to be sure to serve all the shoppers by closing time?

HOMEWORK 1D

1 Without using a calculator, write down the answer to these.

a	50×600	**b**	0.6×40	**c**	0.02×400	**d**	$(30)^2$
e	0.5×250	**f**	0.6×0.7	**g**	$30 \times 40 \times 50$	**h**	$200 \times 0.7 \times 40$

2 Without using a calculator, write down the answer to these.

a	$4000 \div 20$	**b**	$8000 \div 200$	**c**	$400 \div 0.5$	**d**	$2000 \div 0.05$
e	$1800 \div 0.12$	**f**	$600 \div 0.3$	**g**	$200 \times 30 \div 40$	**h**	$300 \times 70 \div 0.4$

AU 3 You are given that $18 \times 21 = 378$

Write down the value of:

a 180×210

b $3780 \div 21$

PS 4 Match each calculation to its answer and then write out the calculations in order, starting with the smallest answer.

6000×300	500×7000	$10\,000 \times 900$	$20 \times 80\,000$
$3\,500\,000$	$1\,800\,000$	$1\,600\,000$	$9\,000\,000$

5 The Moon is approximately 400 000 km from Earth.

If a spaceship takes 8 days to reach the Moon and return, how far does it travel each day?

HOMEWORK 1E

1 Find approximate answers to the following sums.

a	4324×6.71	**b**	6170×7.311	**c**	72.35×3.142
d	4709×3.81	**e**	$63.1 \times 4.18 \times 8.32$	**f**	$320 \times 6.95 \times 0.98$
g	$454 \div 89.3$	**h**	$26.8 \div 2.97$	**i**	$4964 \div 7.23$
j	$316 \div 3.87$	**k**	$2489 \div 48.58$	**l**	$63.94 \div 8.302$

2 Find the approximate monthly pay of the following people whose annual salary is:

a Joy £47 200   **b** Amy £24 200   **c** Tom £19 135

3 Find the approximate annual pay of these brothers who earn:

a Trevor £570 a week   **b** Brian £2728 a month

AU 4 A groundsman bought 350 kg of seed at a cost of £3.84 per kg. Find the approximate total cost of this seed.

FM 5 A greengrocer sells a box of 250 apples for £47. He knows that if he has sold them for 20p each or more he will make a profit.

Did he make a profit? Explain why using approximations.

6 Keith runs about 15 km every day. Approximately how far does he run in:

a a week   **b** a month   **c** a year?

7 A litre of creosote will cover an area of about 6.8 m². Approximately how many litre cans will I need to buy to creosote a fence with a total surface area of 43 m²?

PS 8 A tour of London sets off at 10.13 am and costs £21. It returns at 12.08 pm. Approximately how much is the tour per hour?

HOMEWORK 1F

FM 1 Round each of the following to give sensible answers.
  a Kris is 1.6248 m tall.
  b It took me 17 minutes 48.78 seconds to cook the dinner.
  c My rabbit weighs 2.867 kg.
  d The temperature at the bottom of the ocean is 1.239 °C.
  e There were 23 736 people at the game yesterday.

2 How many jars each holding 119 cm³ of water can be filled from a 3-litre flask?

AU 3 If I walk at an average speed of 62 m per minute, how long will it take me to walk a distance of 4 km?

FM 4 Helen earns £31 500 a year. She works 5 days a week for 45 weeks a year. How much does she earn a day?

5 10 g of gold costs £2.17. How much will 1 kg of gold cost?

FM 6 Rewrite the following article using sensible numbers.
I left home at eleven and a half minutes past two, and walked for 49 minutes. The temperature was 12.7623 °C. I could see an aeroplane overhead at 2937.1 feet. Altogether I had walked 3.126 miles.

PS 7 David travelled 350 miles in 5 hours 10 minutes.
Trevor travelled half the distance in half the time.
Approximately how fast was Trevor travelling?

1.4 Multiples, factors, prime numbers, powers and roots

HOMEWORK 1G

1 From this list of numbers:
  28 19 36 43 64 53 77 66 56 60 15 29 61 45 51

Write down those that are:
  a multiples of 4   **b** multiples of 5   **c** prime numbers   **d** factors of 2700

PS 2 During the peak travel time at a railway station, there are trains setting off to the north every 8 minutes, and there are trains to the south every 12 minutes. At 5 pm one train sets off to the north and one sets off to the south. How many more times will two trains be setting off at the same time before 6.30 pm?

3 Write down the negative square root of each of these.
  a 36   **b** 81   **c** 100   **d** 900   **e** 361
  f 169   **g** 225   **h** 1 000 000 **i** 441   **j** 1225

4 Write down the cube root of each of these.
  a 8   **b** 64   **c** 125   **d** 1000   **e** 27 000
  f −27   **g** −1   **h** −216   **i** −8000   **j** −343

D

AU 5 Here are four numbers.

8   20   25   64

Copy and complete this table by putting the numbers in the correct box.

	Square number	Factor of 40
Cube number		
Multiple of 5		

PS 6 Use these four number cards to make a cube number.

1   **2**   **7**   **9**

AU 7 A number is a factor of 18 and a multiple of 18.
What is the number?

8 Write down the value of each of these.
  a $\sqrt{0.36}$ **b** $\sqrt{0.81}$ **c** $\sqrt{1.69}$ **d** $\sqrt{0.09}$ **e** $\sqrt{0.01}$
  f $\sqrt{1.44}$ **g** $\sqrt{2.25}$ **h** $\sqrt{1.96}$ **i** $\sqrt{4.41}$ **j** $\sqrt{12.25}$

1.5 Prime factors, LCM and HCF

B

HOMEWORK 1H

C

1 Draw factor trees for the following numbers.
  a 144   **b** 75   **c** 98   **d** 420   **e** 560

2 Write the factors for the following in index notation.
  a 48   **b** 54   **c** 216   **d** 1000   **e** 675

AU 3 **a** Express 36 as a product of prime factors.
    b Write your answer to part a in index form.
    c Use you answer to part b to write 18 and 72 as a product of prime factors in index form.

PS 4 $119 = 7 \times 17$
$119^2 = 14\,161$
$119^3 = 1\,685\,159$
  a Write 14 161 as a product of prime factors in index form.
  b Write 1 685 159 as a product of prime factors in index form.
  c Write 119^{10} as a product of prime factors in index form.

PS 5 A mathematician wants to share £18 between three charities so that they each receive a whole number of pounds, and the amount given to each charity is a factor of 18.
How much does he give to each charity?

PS 6 The first three odd prime numbers are all factors of 105.
Explain why this means that seven people can share £105 equally so that each receives an exact number of pounds.

HOMEWORK 1I

1 Find the LCM of each pair of numbers.
  a 5 and 7 **b** 3 and 8 **c** 6 and 9
  d 10 and 12 **e** 10 and 15 **f** 12 and 16
  g 16 and 24 **h** 15 and 35

2 Find the HCF of each pair of numbers.
  a 21 and 49 **b** 27 and 45 **c** 15 and 25
  d 25 and 45 **e** 48 and 60 **f** 72 and 108
  g 54 and 126 **h** 99 and 132

3 Write each of the following as a single power of x.
  a $x^2 \times x^3$ **b** $x^4 \times x^5$ **c** $x^6 \times x$ **d** $x^5 \times x^5$ **e** $x^3 \times x^2 \times x^4$

4 Find the HCF of 55 555 and 67 750.

5 Find the LCM of 144 and 162.

FM 6 Nuts are in packs of 12.
  Bolts are in packs of 18.
  What is the least number of each pack that needs to be bought in order to have the same numbers of nuts and bolts?

PS AU 7 The HCF of two numbers is 5.
  The LCM of the same two numbers is 150.
  What are the numbers?

1.6 Negative numbers

HOMEWORK 1J

FM AU 1 **a** Work out $17 \times (-4)$
  b The average temperature drops by 4°C every day for 17 days. How much has the temperature dropped altogether?
  c The temperature drops by 6°C for the next four days. Write down the calculation to work out the total drop in temperature over these three days.

2 Write down the answers to the following.
  a -2×4 **b** -3×6 **c** -5×7 **d** $-3 \times (-4)$ **e** $-8 \times (-2)$
  f $-14 \div (-2)$ **g** $-16 \div (-4)$ **h** $25 \div (-5)$ **i** $-16 \div (-8)$ **j** $-8 \div (-4)$
  k $3 \times (-7)$ **l** $6 \times (-3)$ **m** $7 \times (-4)$ **n** $-3 \times (-9)$ **o** $-7 \times (-2)$
  p $28 \div (-4)$ **q** $12 \div (-3)$ **r** $-40 \div 8$ **s** $-15 \div (-3)$ **t** $50 \div (-2)$
  u $-3 \times (-8)$ **v** $42 \div (-6)$ **w** $7 \times (-9)$ **x** $-24 \div (-4)$ **y** -7×8

3 Write down the answers to the following.
  a $-2 + 4$ **b** $-3 + 6$ **c** $-5 + 7$ **d** $-3 + (-4)$ **e** $-8 + (-2)$
  f $-14 - (-2)$ **g** $-16 - (-4)$ **h** $25 - (-5)$ **i** $-16 - (-8)$ **j** $-8 - (-4)$
  k $3 + (-7)$ **l** $6 + (-3)$ **m** $7 + (-4)$ **n** $-3 + (-9)$ **o** $-7 + (-2)$
  p $28 - (-4)$ **q** $12 - (-3)$ **r** $-40 - 8$ **s** $-15 - (-3)$ **t** $50 - (-2)$
  u $-3 + (-8)$ **v** $42 - (-6)$ **w** $7 + (8)$ **x** $-24 - (-4)$ **y** $-7 + 8$

4 What number do you multiply -5 by to get the following?
  a 25 **b** -30 **c** 50 **d** -100 **e** 75

PS 5   Put these calculations in order from lowest to highest.

$$-18 \div 12 \qquad -0.5 \times (-4) \qquad -21 \div (-14) \qquad 0.3 \times (-2)$$

HOMEWORK 1K

1   Work out each of these. Remember: first work out the bracket.

   a $-3 \times (-2 + 6)$ **b** $8 \div (-3 + 2)$ **c** $(6 - 8) \times (-3)$ **d** $-4 \times (-6 - 3)$

   e $-5 \times (-6 \div 2)$ **f** $(-5 + 3) \times (-3)$ **g** $(6 - 9) \times (-4)$ **h** $(2 - 5) \times (5 - 2)$

2   Work out each of these.

   a $-5 \times (-4) + 3$ **b** $-8 \div 8 - 3$ **c** $16 \div (-4) + 3$ **d** $2 \times (-5) + 6$

   e $-3 \times 4 - 5$ **f** $-1 + 4^2 - 5$ **g** $5 - 3^2 + 2$ **h** $-1 + 2 \times (-3)$

PS 3   Copy each of these and then put in a bracket to make each one true.

   a $4 \times (-3) + 2 = -4$ **b** $-6 \div (-3) + 2 = 4$ **c** $-6 \div (-3) + 2 = 6$

4 $a = -3, b = 5, c = -4$

   Work out the values of the following.

   a $(a + b)^2$ **b** $-(a + c)^2$ **c** $(a + b)c$ **d** $a^2 + b^2 + c^2$

5   Work out each of the following.

   a $(7^2 + 2^2) \times 3$

   b $18 \div (2 - 5)^2$

   c $3 \times (6^2 - (1 - 8)^2)$

   d $((7 + 1)^2 - (2 - 3)^2) \div 7$

AU 6   Use each of the numbers 4, 6 and 8 and each of the symbols $-$, $\times$ and $\div$ to make a calculation with an answer -3.

AU 7   Use any four different numbers to make a calculation with an answer of -9.

PS 8   Use the numbers 1, 2, 3, 4 and 5, in order from smallest to largest, together with one of each of the symbols $+$, $-$, $\times$, $\div$ and two pairs of brackets to make a calculation with an answer of -2.75.

   For example: Making a calculation with an answer of -4:

$$(1 + 2) - (3 \times 4) + 5 = -4$$

Functional Maths Activity

Flooring specialist

Imagine you are a flooring specialist.

A customer asks you to quote for laying a new floor in her dining room. The room is 5.3 m long and 4.5 m wide. She would prefer luxury carpet or wood, but would consider other types you recommend.

The prices you charge for various types of flooring are given in the table below, along with other useful information.

Flooring type	Size	Pack size (minimum quantity)	Price	Labour cost to lay the flooring
Carpet tiles	500 mm × 500 mm	Packs of 10	£1.89 per pack	£3.50 per m^2
Plain carpet	4 m wide roll	1 m units	£1.39 per m	£4.00 per m^2
Luxury carpet	4 m wide roll	1 m units	£2.99 per m	£4.00 per m^2
Wood (beech)	1.8 m lengths, 200 mm wide	Packs of 5	£15.30 per pack	£8.50 per m^2
Wood (oak)	1.8 m lengths, 200 mm wide	Packs of 5	£22.50 per pack	£8.50 per m^2
Ceramic tiles	200 mm × 300 mm	Packs of 10	£3.49 per pack	£6.50 per m^2

1   Draw up a costing table so that you can tell the customer how much each flooring type would cost. For each flooring type your table needs to show:

- The amount of flooring material required, bearing in mind the minimum purchase quantities
- The cost of material
- The total labour cost to lay the floor
- The total cost to lay the floor
- The cost per metre to lay the floor, including material cost and labour cost
- The area of material (in m^2) that would be wasted (this should be a minimum)
- The cost of the wasted material
- The fraction of the material cost that is wasted (show this as a percentage).

2   The customer has a budget of £350. Can she afford one of her preferred flooring types?

3   What are the advantages and disadvantages of each flooring type? Consider durability, cost, cleaning and any other properties you can think of.

4   Which would you recommend to the customer and why?

2 Number: Fractions and percentages

2.1 One quantity as a fraction of another

1 Write the first quantity as a fraction of the second.
  a 5p, 20p
  b 4 kg, 12 kg
  c 6 hours, 12 hours
  d 14 days, 30 days
  e 6 days, 2 weeks
  f 20 minutes, 2 hours

2 David scored 18 goals out of his team's season total of 48 goals. What fraction of all the goals did David score?

3 James works for 32 weeks a year. What fraction of the year does he work for?

FM 4 Mark earns £120 and saves £40 of it.
Bev earns £150 and saves £60 of it.
Who is saving the greater proportion of their earnings?

FM 5 In a test, Kevin scores 7 out of 10 and Sally scores 15 out of 20. Which is the better mark?
Explain your answer.

AU 6 In a dancing group, there are 20 dancers, of whom 12 are women. Half of the men are single.
What fraction of the dancers are single men?
Give your answer in its simplest form.

PS 7 A bus driver says that at least two out of every three passengers on her bus have a travel pass.
A different driver says it is less than three out of four.
If both statements are true and the bus is carrying 50 passengers, how many have a travel pass? Write down all possible answers.

2.2 Adding and subtracting fractions

1 Evaluate the following.
  a $\frac{1}{2} + \frac{1}{5}$
  b $\frac{1}{2} + \frac{1}{3}$
  c $\frac{1}{3} + \frac{1}{10}$
  d $\frac{3}{8} + \frac{1}{3}$
  e $\frac{3}{4} + \frac{1}{5}$
  f $\frac{1}{3} + \frac{2}{5}$
  g $\frac{3}{5} + \frac{3}{8}$
  h $\frac{1}{2} + \frac{2}{5}$

2 Evaluate the following.
  a $\frac{1}{2} + \frac{1}{4}$
  b $\frac{1}{3} + \frac{1}{6}$
  c $\frac{3}{5} + \frac{1}{10}$
  d $\frac{5}{8} + \frac{1}{4}$

3 Evaluate the following.
  a $\frac{7}{8} - \frac{3}{4}$
  b $\frac{4}{5} - \frac{1}{2}$
  c $\frac{2}{3} - \frac{1}{5}$
  d $\frac{3}{4} - \frac{2}{5}$

4 Evaluate the following.
  a $\frac{5}{8} + \frac{3}{4}$
  b $\frac{1}{2} + \frac{3}{5}$
  c $\frac{5}{6} + \frac{1}{4}$
  d $\frac{2}{3} + \frac{3}{4}$

FM Functional Maths **AU** (AO2) Assessing Understanding **PS** (AO3) Problem Solving

AU 5 **a** At a football club, half of the players are English, a quarter are Scottish and one-sixth are Italian. The rest are Irish. What fraction of players at the club are Irish?

b One of the following is the number of players at the club:

      30   32   34   36

How many players are at the club?

6 On a firm's coach trip, half the people were employees and two-fifths were partners of the employees. The rest were children. What fraction of the people were children?

7 Five-eighths of a crowd of 35 000 people were male. How many females were in the crowd?

8 What is four-fifths of 65 added to five-sixths of 54?

PS 9 Here are four fractions.

$$\frac{1}{6} \qquad \frac{5}{12} \qquad \frac{1}{4} \qquad \frac{1}{3}$$

Which three of them add up to 1?

PS FM 10 Pipes are made in $\frac{1}{2}$ m and $\frac{3}{4}$ m lengths.

Show how it is possible to make a pipe exactly 2 m long using the least number of $\frac{1}{2}$ m and $\frac{3}{4}$ m pipes.

Assume that you cannot cut pipes to size.

2.3 Multiplying fractions

HOMEWORK 2C

1 Evaluate the following, leaving your answer in its simplest form.

a $\frac{1}{2} \times \frac{2}{3}$ **b** $\frac{3}{4} \times \frac{2}{5}$ **c** $\frac{3}{5} \times \frac{1}{2}$ **d** $\frac{3}{7} \times \frac{2}{3}$ **e** $\frac{2}{3} \times \frac{5}{6}$

f $\frac{1}{3} \times \frac{3}{5}$ **g** $\frac{2}{3} \times \frac{7}{10}$ **h** $\frac{3}{8} \times \frac{2}{5}$ **i** $\frac{4}{9} \times \frac{3}{8}$ **j** $\frac{4}{5} \times \frac{7}{16}$

2 Kris walked three-quarters of the way along Carterknowle Road which is 3 km long. How far did Kris walk?

AU 3 Jean ate one-fifth of a cake, Les ate a half of what was left. Nick ate the rest. What fraction of the cake did Nick eat?

4 A formula for working out the distance travelled is:

    Distance travelled = Speed × Time taken

A snail is moving at one-tenth of a metre per minute. It travels for half a minute. How far has it travelled?

PS 5 George is given £80.

Each week he spends half of the amount left.

What fraction of the £80 will he have left after 4 weeks?

AU 6 You are given that 1 litre = 100 cl.

A bottle holds 1.5 litres.

Tupac drinks half of the contents.

Belinda drinks 25 cl of the contents.

What fraction of the contents are left?

7 Evaluate the following, leaving your answer as a mixed number where possible.

a $1\frac{1}{3} \times 2\frac{1}{4}$ **b** $1\frac{3}{4} \times 1\frac{1}{3}$ **c** $2\frac{1}{2} \times \frac{4}{5}$ **d** $1\frac{2}{3} \times 1\frac{3}{10}$

e $3\frac{1}{4} \times 1\frac{3}{5}$ **f** $2\frac{2}{3} \times 1\frac{3}{4}$ **g** $3\frac{1}{2} \times 1\frac{1}{6}$ **h** $7\frac{1}{2} \times 1\frac{3}{5}$

PS 8 Which is the smaller, $\frac{3}{4}$ of $5\frac{1}{3}$ or $\frac{2}{3}$ of $4\frac{2}{5}$?

AU 9 I estimate that I need 60 litres of lemonade for a party.
I buy 24 bottles, each containing $2\frac{3}{4}$ litres.
Have I bought enough lemonade for the party?

PS 10 Pizzas are often cut into 8 equal pieces.
If a pizza is cut into six equal pieces how much more pizza is in each piece?
Give your answer as a fraction of the whole pizza.

AU 11 A company employs 200 people.
The manager says that exactly two-thirds of the employees are women and three-quarters of the employees are full-time.
One of the statements is true and one is not accurate.
Explain which statement is which.

FM 12 A fish farmer is trying to work out how many fish are in a pond.
He captures 100 fish, marks them and puts them back in the pond.
Later he captures 100 fish and finds that 25 are marked.
Approximately how many fish are in the pond?

2.4 Dividing by a fraction

HOMEWORK 2D

1 Evaluate the following, leaving your answer as a mixed number where possible.
a $\frac{1}{5} \div \frac{1}{3}$ b $\frac{3}{5} \div \frac{3}{8}$ c $\frac{4}{5} \div \frac{2}{3}$ d $\frac{4}{7} \div \frac{8}{9}$
e $4 \div 1\frac{1}{2}$ f $5 \div 3\frac{2}{3}$ g $8 \div 1\frac{3}{4}$ h $6 \div 1\frac{1}{4}$
i $5\frac{1}{2} \div 1\frac{1}{3}$ j $7\frac{1}{2} \div 2\frac{2}{3}$ k $1\frac{1}{2} \div 1\frac{1}{5}$ l $3\frac{1}{5} \div 3\frac{3}{4}$

FM 2 A pet shop has 36 kg of hamster food. Tom, who owns the shop, wants to pack this into bags, each containing three-quarters of a kilogram. How many bags can he make in this way?

AU 3 Bob is putting a fence down the side of his garden, it is to be 20 m long. The fence comes in sections; each one is $1\frac{1}{3}$ m long. How many sections will Bob need to put the fence all the way down the one side of his garden?

PS 4 An African bullfrog can jump a distance of $1\frac{1}{4}$ m in one hop. How many hops would it take an African bullfrog to hop a distance of 100 m?

AU 5 Three-fifths of all 14-year-olds in a school visit the dentist each year.
One-third of those who do not visit the dentist have a problem with their teeth.
What fraction of the 14-year-olds do not visit the dentist and have a problem with their teeth?

FM 6 How many half-litre tins of paint can be poured into a 2.2 litre paint tray without spilling?

PS 7 I work 8 hours in a day.
Short tasks last $\frac{1}{4}$ hour.
Long tasks last $\frac{3}{4}$ hour.
I have to complete at least three long tasks.
How many short tasks can I also do in one working day?

8 Evaluate the following, leaving your answer as a mixed number wherever possible.

a $\frac{4}{5} \times \frac{1}{2} \times \frac{3}{8}$ **b** $\frac{3}{4} \times \frac{7}{10} \times \frac{5}{6}$ **c** $\frac{2}{3} \times \frac{5}{6} \times \frac{9}{10}$

d $1\frac{1}{4} \times \frac{2}{3} \div \frac{5}{6}$ **e** $\frac{5}{8} \times 1\frac{1}{3} \div 1\frac{1}{10}$ **f** $2\frac{1}{2} \times 1\frac{1}{3} \div 3\frac{1}{3}$

2.5 Increasing and decreasing quantities by a percentage

HOMEWORK 2E

1 Increase each of the following by the given amount.

     a £80 by 5%   **b** 14 kg by 6%   **c** £42 by 3%

2 Increase each of the following by the given amount.

     a 340 g by 10%   **b** 64 m by 5%   **c** £41 by 20%

3 Keith, who was on a salary of £34 200, was given a pay rise of 4%. What was his new salary?

4 In 1991 the population of Dripfield was 14 200. By 2001 that had increased by 8%. What was the population of Dripfield in 2001?

5 In 1993 the number of bikes on the roads of Doncaster was about 840. Since then it has increased by 8%. Approximately how many bikes are on the roads of Doncaster now?

PS 6 A dining table costs £300 before the VAT is added.
If the rate of VAT goes up from 15% to 20%, how much will the cost of the dining table increase?

FM 7 A restaurant meal is advertised at £20 (a service charge will be added to all bills).

Romano's Bistro	
2 × Prawn Cocktail	4.60
1 × Chicken Risotto	6.60
1 × Sea Bass	8.50
1 × Fries	4.50
1 × Tiramisu	6.60
1 × Choc fudge cake	5.20
2 × Coffee	5.00
Service charge	5.00
Total	46.00

The bill for two people is shown.
Show that the service charge is 15%.

PS 8 A shopkeeper decides to increase prices by 5% or £1, whichever is greater.
What is the price range of the items that will rise by £1?

HOMEWORK 2F

1 Decrease each of the following by the given percentage. (Use any method you like.)

     a £20 by 10%   **b** £150 by 20%   **c** 90 kg by 30%   **d** 500 m by 12%

     e £260 by 5%   **f** 80 cm by 25%   **g** 400 g by 42%   **h** £425 by 23%

     i 48 kg by 75%   **j** £63 by 37%

FM 2 Mrs Denghali buys a new car from a garage for £8400. The garage owner tells her that the value of the car will lose 24% after one year. What will be the value of the car after one year?

3 The population of a village in 2006 was 2400. In 2010 the population had decreased by 12%. What was the population of the village in 2010?

FM 4 A Travel Agent is offering a 15% discount on holidays. How much will the advertised holiday now cost?

NEW YORK FOR A WEEK
£540

FM 5

New Year's Sale:
All prices reduced by 20%

Matt was given £160 at Christmas. Can he afford to buy a shirt that normally costs £30, a suit that normally costs £130 and a pair of shoes that normally cost £42?

6 On the first day of a new term, a school expects to have an attendance rate of 99%. If the school population is 700 pupils, how many pupils will the school expect to be absent on the first day of the new term?

FM 7 By putting cavity wall insulation into your home, you could use 20% less fuel. A family using an average of 850 units of electricity a year put cavity wall insulation into their home. How much electricity would they expect to use now?

AU PS 8 A shop increases all its prices by 10%.

One month later it advertises 10% off all marked prices.
Are the goods cheaper, the same or more expensive than before the price increase?
Show how you work out your answer.

PS 9 Prove that a 10% increase followed by a 10% decrease is equivalent to a 1% increase overall.

2.6 Expressing one quantity as a percentage of another

HOMEWORK 2G

1 Express each of the following as a percentage. Give your answers to one decimal place where necessary.
  a £8 of £40   **b** 20 kg of 80 kg   **c** 5 m of 50 m
  d £15 of £20   **e** 400 g of 500 g   **f** 23 cm of 50 cm
  g £12 of £36   **h** 18 minutes of 1 hour   **i** £27 of £40
  j 5 days of 3 weeks

PS 2 What percentage of these shapes is shaded?
  a   **b**

3 In a class of 30 pupils, 18 are girls.

    a What percentage of the class are girls?

    b What percentage of the class are boys?

4 The area of a farm is 820 hectares. The farmer uses 240 hectares for pasture.

What percentage of the farm land is used for pasture? Give your answer to one decimal place.

5 Find the percentage profit on the following items. Give your answers to one decimal place.

Item (Selling price)	Retail price (Price the shop paid)	Wholesale price
Micro hi-fi system	£250	£150
CD radio cassette	£90	£60
MiniDisc player	£44.99	£30
Cordless headphones	£29.99	£18

AU 6 Paul and Val take the same tests. Both tests are worth the same number of marks.
Here are their results.

	Test A	Test B
Paul	30	40
Val	28	39

Whose result has the greater percentage increase from test A to test B?
Show your working.

PS 7 A small train is carrying 48 passengers.
At a station, more passengers get on so that all seats are filled and no one is standing.
At the next station, 70% of the passengers leave the train and 30 new passengers get on.
There are now 48 passengers on the train again.
How many seats are on the train?

PS 8 In a secondary school, 30% of students have a younger brother or sister at primary school.
20% of students have two younger brothers or sisters at primary school
Altogether there are 700 brothers and sisters at the primary school.
How many students are at the secondary school?

FM 9 James came home from school one day with his end-of-year test results. Change each of James' results to a percentage.

Maths   63 out of 75   English   56 out of 80

Science   75 out of 120   French   27 out of 60

10 During the wet year of 1993, it rained in London on 80 days of the year. What percentage of the days of the year were wet? Give your answer to two significant figures.

2.7 Compound interest and repeated percentage change

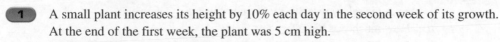

HOMEWORK 2H

1 A small plant increases its height by 10% each day in the second week of its growth.
At the end of the first week, the plant was 5 cm high.
What is its height after a further:

    a 1 day   **b** 2 days   **c** 4 days   **d** 1 week?

FM 2 The headmaster of a new school offered his staff an annual pay increase of 5% for every year they stayed with the school.

a Mr Speed started teaching at the school on a salary of £28 000. What salary will he be on after 3 years if he stays at the school?

b Miss Tuck started teaching at the school on a salary of £14 500. How many years will it be until she is earning a salary of over £20 000?

3 Billy put a gift of £250 into a special savings account that offered him 8% compound interest if he promised to keep the money in for at least 2 years. How much was in this account after:

a 2 years **b** 3 years **c** 5 years?

4 The penguin population of a small island was only 1500 in 1998, but it steadily increased by about 15% each year. Calculate the population in:

a 1999 **b** 2000 **c** 2002

PS 5 A sycamore tree is 40 cm tall; it grows at a rate of 8% per year. A conifer is 20 cm tall. It grows at a rate of 15% per year. How many years does it take before the conifer is taller than the sycamore?

AU 6 The population of a small town is 2000.
The population is falling by 10% each year.
The population of a nearby village is 1500.
This population is rising by 10% each year.
After how many years will the population of the town be less than the population of the village?

PS 7 Each week, a boy takes out 20% of the amount in his bank account to spend.
After how many weeks will the amount in his bank account have halved from the original amount?

2.8 Reverse percentage (working out the original quantity)

HOMEWORK 2I

1 Find what 100% represents when:
a 20% represents 160 g **b** 25% represents 24 m **c** 5% represents 42 cm.

2 Find what 100% represents when:
a 40% represents 28 kg **b** 30% represents £54 **c** 15% represents 6 hours.

 3 VAT is a government tax added to goods and services. With VAT at 17.5%, what is the pre-VAT price of the following priced goods?
Jumper £14.10   Socks £1.88   Trousers £23.50

 **4** Paula spends £9 each week on CDs. This is 60% of her weekly income. How much is Paula's weekly income?

5 Alan's weekly pay is increased by 4% to £187.20. What was Alan's pay before the increase?

 FM 6 Jon's salary is now £23 980. This is 10% more than he earned two years ago.
Last year his salary was 3% more than it was two years ago.
a How much was his salary last year?
b By what percentage was his salary increased last year?

7 Twice as many people visit a shopping centre on Saturdays than on Fridays.

The number visiting on both days increases by 50% in the week before Christmas. How many more visit on this Saturday than on this Friday? Give your answer as a percentage.

PS 8 A man's savings decreased by 10% in one year and then increased in the following year by 10%.
He now has £1782.
How much did he have two years ago?

Functional Maths Activity

Value Added Tax

You are a shopkeeper and need to use a table to work out the VAT you charge on items for sale.

VAT table

Rate	£1	£10	£100	£500	£1000	£2000	£5000	£10 000
5%	5p	50p	£5					
8%				£40				
15%					£150			
17.5%						£350		
20%							£1000	

Total cost table

Rate	£1	£10	£100	£500	£1000	£2000	£5000	£10 000
5%	£1.05	£10.50	£105					
8%				£540				
15%					£1150			
17.5%						£350		
20%							£6000	

Task 1
Complete the tables.

Task 2

All Prices Exclusive of VAT

Here are some items from the shop priced excluding VAT.
Use the tables to help you work out the price including VAT at 17.5%.

Functional Maths Activity (continued)

Task 3

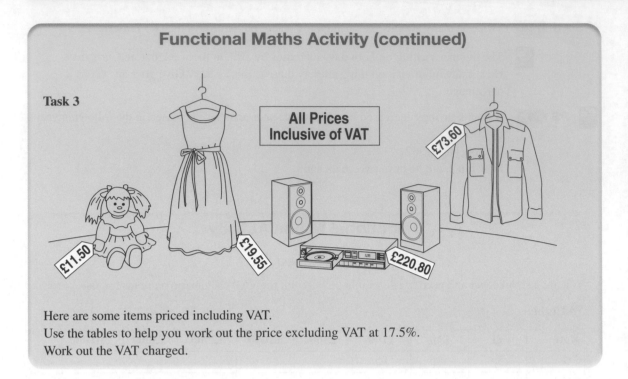

All Prices
Inclusive of VAT

£11.50

£19.55

£73.60

£220.80

Here are some items priced including VAT.
Use the tables to help you work out the price excluding VAT at 17.5%.
Work out the VAT charged.

Number: Ratios and proportion

3.1 Ratio

1 Express each of the following ratios in their simplest form.

a 3 : 9   **b** 5 : 25   **c** 4 : 24   **d** 10 : 30   **e** 6 : 9

f 12 : 20   **g** 25 : 40   **h** 30 : 4   **i** 14 : 35   **j** 125 : 50

2 Express each of the following ratios of quantities in their simplest form. (Remember to change to a common unit where necessary.)

a £2 to £8   **b** £12 to £16   **c** 25 g to 200 g

d 6 miles : 15 miles   **e** 20 cm : 50 cm   **f** 80p : £1.50

g 1 kg : 300g   **h** 40 seconds : 2 minutes   **i** 9 hours : 1 day

j 4 mm : 2 cm

3 £20 is shared out between Bob and Kathryn in the ratio 1 : 3.

a What fraction of the £20 does Bob receive?

b What fraction of the £20 does Kathryn receive?

4 In a class of students, the ratio of boys to girls is 2 : 3.

a What fraction of the class is boys?

b What fraction of the class is girls?

FM 5 Pewter is an alloy containing lead and tin in the ratio 1 : 9.

a What fraction of pewter is lead?

b What fraction of pewter is tin?

AU 6 Roy wins two-thirds of his snooker matches. He loses the rest.

What is his ratio of wins to losses?

PS 7 In the 2009 Ashes cricket series, the numbers of wickets taken by Steve Harmison and Monty Panesar were in the ratio 5 : 1.

The ratio of the number of wickets taken by Graham Onions to those taken by Steve Harmison was 2 : 1.

What fraction of the wickets taken by these three bowlers was by Monty Panesar?

FM 8 In a motor sales business, the ratio of the area allocated to selling new cars to the area for selling used cars is 3 : 2.

Half the area for selling new cars is changed to used cars.

What is the ratio of the areas now?

Give your answer in its simplest form.

1 Divide each of the following amounts in the given ratios.

a £10 in the ratio 1 : 4   **b** £12 in the ratio 1 : 2

c £40 in the ratio 1 : 3   **d** 60 g in the ratio 1 : 5

e 10 hours in the ratio 1 : 9

FM Functional Maths   **AU** (AO2) Assessing Understanding   **PS** (AO3) Problem Solving

2 The ratio of female to male members of a sports centre is 3 : 1. The total number of members of the centre is 400.
  a How many members are female?   **b** How many members are male?

3 A 20 m length of cloth is cut into two pieces in the ratio 1 : 9. How long is each piece?

4 Divide each of the following amounts in the given ratios.
  a 25 kg in the ratio 2 : 3   **b** 30 days in the ratio 3 : 2
  c 70 m in the ratio 3 : 4   **d** £5 in the ratio 3 : 7
  e 1 day in the ratio 5 : 3

5 James collects beer mats and the ratio of British mats to foreign mats in his collection is 5 : 2. He has 1400 beer mats. How many foreign beer mats does he have?

6 Patrick and Jane share out a box of sweets in the ratio of their ages. Patrick is 9 years old and Jane is 11 years old. If there are 100 sweets in the box, how many does Patrick get?

AU 7 For her birthday Reena is given £30. She decides to spend four times as much as she saves. How much does she save?

8 Mrs Megson calculates that her quarterly electric and gas bills are in the ratio 5 : 6. The total she pays for both bills is £66. How much is each bill?

9 You can simplify a ratio by changing it into the form 1 : n. For example, 5 : 7 can be rewritten as 5 : 7 = 1 : 1.4 by dividing each side of the ratio by 5. Rewrite each of the following ratios in the form 1 : n.
  a 2 : 3   **b** 2 : 5   **c** 4 : 5   **d** 5 : 8   **e** 10 : 21

PS 10 The amount of petrol and diesel sold at a garage is in the ratio 2 : 1. One-tenth of the diesel sold is bio-diesel.
What fraction of all the fuel sold is bio-diesel?

PS FM 11 At a buffet, there are twice as many men as women.
The organiser takes £600 altogether.
There are 50 men at the buffet.
How much does each person pay for the buffet?

FM AU 12 The cost of show tickets for adults and children are in the ratio 5 : 3.
30 adults and 40 children visit the show.
Children's tickets cost £3.60.
How much money will the show take altogether?

HOMEWORK 3C

1 Peter and Margaret's ages are in the ratio 4 : 5. If Peter is 16 years old, how old is Margaret?

2 Cans of lemonade and packets of crisps were bought for the school disco in the ratio 3 : 2. The organiser bought 120 cans of lemonade. How many packets of crisps did she buy?

FM 3 In his restaurant, Manuel is making fruit punch, a drink made from fruit juice and iced soda water, mixed in the ratio 2 : 3. Manuel uses 10 litres of fruit juice.
  a How many litres of soda water does he use?
  b How many litres of fruit punch does he make?

4 Cupro-nickel coins are minted by mixing copper and nickel in the ratio 4 : 1.
  a How much copper is needed to mix with 20 kg of nickel?
  b How much nickel is needed to mix with 20 kg of copper?

5 The ratio of male to female spectators at a school inter-form football match is 2 : 1. If 60 boys watched the game, how many spectators were there in total?

6 Marmalade is made from sugar and oranges in the ratio 3 : 5. A jar of 'Savilles' marmalade contains 120 g of sugar.
  a How many grams of oranges are in the jar?
  b How many grams of marmalade are in the jar?

AU 7 Each year Abbey School holds a sponsored walk for charity. The money raised is shared between a local charity and a national charity in the ratio 1 : 2. Last year the school gave £2000 to the local charity.
  a How much did the school give to the national charity?
  b How much did the school raise in total?

PS 8 Fred's blackcurrant juice is made from 4 parts blackcurrant and 1 part water. Jodie's blackcurrant juice is made from blackcurrant and water in the ratio 7 : 2. Which juice contains the greater proportion of blackcurrant? Show how you work out your answer.

9 Sand and cement is mixed in the ratio 3 : 1. Cement is sold in 25 kg bags. Sand is sold in 875 kg sacks. How many sacks of sand would be needed to mix with 20 bags of cement?

AU 10 The ratio of tins of white paint to coloured paint in a shop storeroom is 2 : 5. There is enough room on the shelves for 60 tins of paint. How many tins of white paint can be put on the shelf if the ratio of the tins of white paint to coloured paint is also 2 : 5?

3.2 Speed, time and distance

HOMEWORK 3D

1 A cyclist travels a distance of 60 miles in 4 hours. What was her average speed?

2 How far along a motorway will you travel if you drive at an average speed of 60 mph for 3 hours?

3 Mr Baylis drives on a business trip from Manchester to London in $4\frac{1}{2}$ hours. The distance he travels is 207 miles. What is his average speed?

4 The distance from Leeds to Birmingham is 125 miles. The train I catch travels at an average speed of 50 mph. If I catch the 11.30 am train from Leeds, at what time would I expect to be in Birmingham?

5 Copy and complete the following table.

	Distance travelled	Time taken	Average speed
a	240 miles	8 hours	
b	150 km	3 hours	
c		4 hours	5 mph
d		$2\frac{1}{2}$ hours	20 km/h
e	1300 miles		400 mph
f	90 km		25 km/h

6 A coach travels at an average speed of 60 km/h for 2 hours on a motorway and then slows down in a town centre to do the last 30 minutes of a journey at an average speed of 20 km/h.

   a What is the total distance of this journey?

   b What is the average speed of the coach over the whole journey?

7 Hilary cycles to work each day. She cycles the first 5 miles at an average speed of 15 mph and then cycles the last mile in 10 minutes.

   a How long does it take Hilary to get to work?

   b What is her average speed for the whole journey?

8 Martha drives home from work in 1 hour 15 minutes. She drives home at an average speed of 36 mph.

   a Change 1 hour 15 minutes to decimal time in hours.

   b How far is it from Martha's work to her home?

PS 9 A tram route takes 15 minutes at an average speed of 16 mph.

The same journey by car is two miles longer.

How fast would a car need to travel to arrive in the same time?

FM 10 A taxi travelled for 30 minutes.

The fare was £24.

If the fare was charged at £1.20 per mile, what was the average speed of the taxi?

AU 11 Two cars are 30 miles apart but travelling towards each other.

The average speed of one car is twice as fast as the other car.

The slower car is averaging 20 mph.

How long is it before they meet up?

3.3 Direct proportion problems

HOMEWORK 3E

1 If four video tapes cost £3.20, what would 10 video tapes cost?

2 Five oranges cost 90p. Find the cost of 12 oranges.

3 Dylan earns £18.60 in 3 hours. How much will he earn in 8 hours?

4 Barbara bought 12 postcards for 3 euros when she was on holiday in Tenerife.

   a How many euros would she have paid if she had only bought 9 postcards?

   b How many postcards could she have bought with a 5 euro note?

FM 5 Five 'Day-Rover' bus tickets cost £8.50.

   a What is the cost of 16 tickets?

   b Pat has £20. She wants to buy 12 tickets.

      Can she afford them?

      Show your working of how you decide.

6 A car uses 8 litres of petrol on a trip of 72 miles.

   a How much would be used on a trip of 54 miles?

   b How far would the car go on a full tank of 45 litres?

7 It takes a photocopier 18 seconds to produce 12 copies. How long will it take to produce 32 copies?

8 Val has a recipe for making 12 flapjacks.
　　100 g margarine
　　4 tablespoons golden syrup
　　80 g granulated sugar
　　200 g rolled oats

FM **a** What is the recipe for:
　　i 6 flapjacks　　**ii** 24 flapjacks　　**iii** 30 flapjacks?

PS **b** What is the maximum number of flapjacks she can make if she has 1 kg of each ingredient?

AU 9 Greg the baker sells bread rolls in pack of 6 for £1. Dom the baker sells bread rolls in packs of 24 for £3.19.
I have £5 to spend on bread rolls.
How many more can I buy from Greg than from Dom?

FM 10 To make white coffee, one-quarter of a cup is filled with milk.
A cup holds 600 ml of white coffee altogether.
How many cups of coffee can be made if you have 1 litre of milk?

PS 11 A nurse can examine 20 patients each hour.
There are 170 patients visiting a three-hour clinic.
How many nurses are needed?

3.4 Best buys

HOMEWORK 3F

FM 1 Compare the prices of the following pairs of products and state which, if any, is the better buy.
a Mouthwash: £1.99 for a twin pack, or £1.49 each with a 3 for 2 offer.
b Dusters: 79p for a pack of 6 with a 'buy one pack and get one pack free' offer, or £1.20 for a pack of 20.

 2 Compare the following pairs of products and state which is the better buy and why.
a Tomato ketchup: a medium bottle which is 200 g for 55p or a large bottle which is 350 g for 87p.
b Milk chocolate: a 125 g bar at 77p or a 200 g bar at 92p.
c Coffee: a 750 g tin at £11.95 or a 500 g tin at £7.85.
d Honey: a large jar which is 900 g for £2.35 or a small jar which is 225 g for 65p.

 3 Boxes of 'Wetherels' teabags are sold in three different sizes.

Small

80 teabags
£1.44

Medium

120 teabags
£2.10

Large

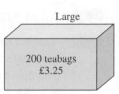

200 teabags
£3.25

Which size box of teabags gives the best value for money?

D

 4 Bottles of 'Cola' are sold in different sizes. Copy and complete the table.

Size of bottle	Price	Cost per litre
$\frac{1}{2}$ litre	36p	
$1\frac{1}{2}$ litres	99p	
2 litres	£1.40	
3 litres	£1.95	

Which size of bottle gives the best value for money?

AU 5 The following 'special offers' were being promoted by a supermarket.

Only £1.99 each   Buy 3 for the price of 2

Cornflakes
750 g
£1.99

Cornflakes
500 g
£1.69

Which offer is the better value for money? Explain why.

PS 6 Hannah scored 17 out of 20 in a test. John scored 40 out of 50 in a test of the same standard.
Who got the better mark?

3.5 Density

 **HOMEWORK 3G**

B

1 Find the density of a piece of wood weighing 135 g and having a volume of 150 cm^3.

2 Calculate the density of a metal if 40 cm^3 of it weighs 2500 g.

3 Calculate the weight of a piece of plastic, 25 cm^3 in volume, if its density is 1.2 g/cm^3.

4 Calculate the volume of a piece of wood that weighs 350 g and has a density of 0.7 g/cm^3.

5 Find the weight of a statue, 540 cm^3 in volume, if the density of marble is 2.5 g/cm^3.

6 Calculate the volume of a liquid weighing 1 kg and having a density of 1.1 g/cm^3.

7 Find the density of a stone which weighs 63 g and has a volume of 12 cm^3.

PS 8 It is estimated that a huge rock balanced on the top of a mountain has a volume of 120 m^3. The density of the rock is 8.3 g/cm^3. What is the estimated weight of the rock?

9 A 1 kg bag of flour has a volume of about 900 cm^3. What is the density of flour in g/cm^3?

AU 10 1 m^3 = 1 000 000 cm^3
A storage area has 30 tonnes of sandstone.
The density of some sandstone is 2.3 g/cm^3
a What is the volume of sandstone in the storage area?
Give your answer in m^3.
b The density of granite is 2.7 g/cm^3.
The same volume of granite is stored as the volume of sandstone.
How much heavier is the granite?

AU 11 The density of a piece of oak is 630 kg/m³.
The density of a piece of mahogany is 550 kg/m³.
Two identical carvings are made, one from oak and the other from mahogany.
The oak carving has a mass of 315 grams.
What is the mass of the mahogany carving?

FM 12 Two identical-looking metal objects are side by side.
They have different masses.
How does this tell you that they are probably made from different metals?

PS 13 Metal A has density that is half the density of metal B.

Functional Maths Activity

Metal objects

You run a charity shop and the items listed in the table below have been donated. You have estimated their volume in cubic centimetres. Now you need to price them. The second table provides data about the density of metal and the cost per gram.

Use the information to work out the value of each item.

How much would you price each item at and why?

Item	Volume (cm³)
Candlestick (brass)	6
Statue (cast iron)	15
Ring (gold)	0.5
Tankard (Stainless steel)	4
Jug (silver)	3
Plate (Copper)	7

Metal	Density (g/cm³)	Cost per gram
Brass	8.5	29p
Copper	8.9	22p
Gold	19.3	£16.58
Silver	10.4	80p
Stainless steel	7.5	33p
Cast iron	7.2	83p

Geometry: Length, area and volume

4.1 Circumference and area of a circle

HOMEWORK 4A

1 Find the circumference of each of the following circles, round off your answers to 1 dp.
  a Diameter 3 cm   **b** Radius 5 cm   **c** Radius 8 m
  d Diameter 14 cm   **e** Diameter 6.4 cm   **f** Radius 3.5 cm

2 John runs twice round a circular track which has a radius of 50 m. How far has he run? Give your answers in terms of π.

3 A rolling pin has a diameter of 5 cm.
  a What is the circumference of the rolling pin?
  b How many revolutions does it make when rolling a length of 30 cm?

4 How many complete revolutions will a bicycle wheel with a radius of 28 cm make in a journey of 3 km?

5 Calculate the area of each of these circles, giving your answers to 1 decimal place, except for **a** and **d**, where your answer should be in terms of π.
  a Radius 4 cm   **b** Diameter 14 cm   **c** Radius 9 cm
  d Diameter 2 m   **e** Radius 21 cm   **f** Diameter 0.9 cm

6 What is the total perimeter of a semicircle of diameter 7 cm? Give your answer to 1 dp.

7 What is the total perimeter of a semicircle of radius 6 cm? Give your answer in terms of π.

8 A circle has a circumference of 12 cm. What is its diameter?

9 A garden has a circular lawn of diameter 20 m. There is a path 1 m wide all the way round the circumference. What is the area of this path?

10 Calculate the area of a semicircle with a diameter of 15 cm. Give your answer to 1 dp.

11 A circle has an area of 50 m². What is its radius?

12 I have a circle with a circumference of 25 cm. What is the area of this circle?

13 Jane walked around a circular lawn. She counted 153 paces to walk round it. Each of her paces was about 42 cm. What is the area of the lawn?

AU 14 Calculate the area of this shape.

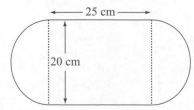

FM Functional Maths **AU** (AO2) Assessing Understanding **PS** (AO3) Problem Solving

FM 15 The diameter of a cycle wheel is 70 cm. How many metres will the cycle travel if the wheel makes 50 revolutions?

4.2 Area of a trapezium

HOMEWORK 4B

 1 Calculate the perimeter and the area of each of these trapeziums.

a

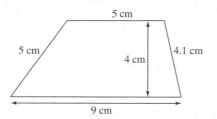

b

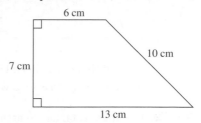

2 Calculate the area of each of these shapes.

a

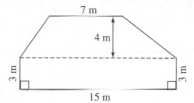

b

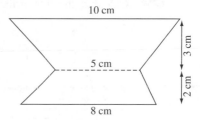

 3 Which of the following shapes has the largest area?

a

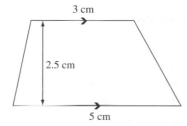

b

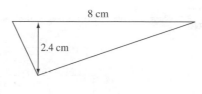

 AU 4 Calculate the area of this trapezium.

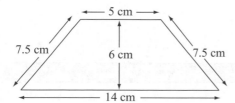

D

 FM 5 This is the plan of an area that is to be seeded with grass.

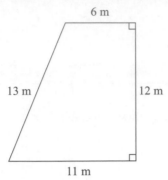

6 m

13 m   12 m

11 m

Seed should be planted at a rate of 30 g per m².
How much grass seed will be required?

PS 6 A trapezium has an area of 100 cm². The parallel sides are 17 cm and 23 cm long.
How far apart are the parallel sides?

7 Calculate the area of the shaded part in each of these diagrams.

C

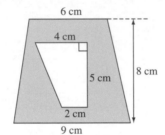 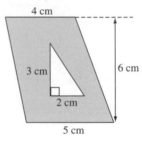

a 6 cm   **b** 4 cm

4 cm   3 cm
5 cm   8 cm   2 cm   6 cm
2 cm
9 cm   5 cm

 8 What percentage of this shape has been shaded?

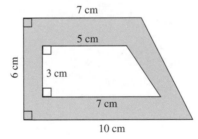

7 cm

5 cm

6 cm   3 cm

7 cm

10 cm

4.3 Sectors

HOMEWORK 4C

A

 1 For these sectors, calculate the arc length and the sector area.

a

50°
10 cm

b

90°
7 cm

2 Calculate the arc length and the area of a sector whose arc subtends a right angle in a
circle of diameter 10 cm. Give your answer in terms of π.

3 Calculate the total perimeter of each of these shapes.

a

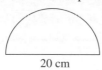

20 cm

b

12 cm

 4 Calculate the area of each of these shapes.

a

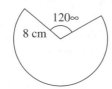

120°
8 cm

b

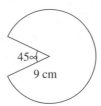

45°
9 cm

 5 There is an infrared sensor in a security system. The sensor can detect movement inside a sector of a circle. The radius of the circle is 16 m. The sector is 120°. Calculate the area of the sector.

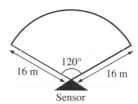

120°
16 m   16 m
Sensor

 AU 6 A circle of radius 8 cm is cut up into five congruent sectors. Calculate the perimeter of each one.

 FM 7 A shelf to fit in the corner of a room is to be cut in the shape of a quarter of a circle. It will be cut from a square of wood of side 30 cm.
What will be the area of the shelf?

PS 8 ABCD is a square of side length 15 cm. APC and AQC are arcs of the circle with centres D and B. Calculate the area of the unshaded part.

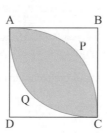

A   B
        P
Q
D   C

4.4 Volume of a prism

HOMEWORK 4D

1 For each prism shown, calculate the area of the cross-section and the volume.

a

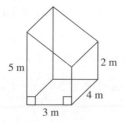

5 m   2 m
        4 m
3 m

b

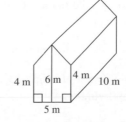

4 m   6 m   4 m   10 m
       5 m

A chocolate box is in the form of a triangular prism. It is 18 cm long and has a volume of 387 cm³.

What is the area of the triangular end of the box?

 A wooden door wedge has a cross-section which is this shape:

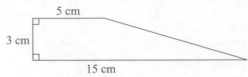

The wedge is 3 cm wide.

a Calculate the volume of wood needed to make the wedge.

b If the wedge is cut from a block of wood (cuboid) measuring 15 cm × 3 cm × 3 cm, what volume of wood is wasted?

  Which of these solids is:

a the heaviest   **b** the lightest?

i (1.26 g/cm³)   **ii**   **iii**

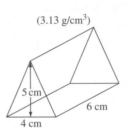 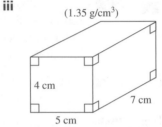

4.5 Cylinders

HOMEWORK 4E

1 Find **i** the volume and **ii** the curved surface area of a cylinder with base radius 5 cm and height 4 cm. Give your answer in terms of π.

2 Find **i** the volume and **ii** the curved surface area of a cylinder with base radius 8 cm and height 17 cm. Give your answer to a suitable degree of accuracy.

3 Find **i** the volume and **ii** the total surface area of each of these cylinders.

a   5 cm   **b**

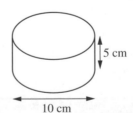

4 What is the radius of a cylinder with a height of 6 cm and a volume of 24π cm³?

5 What is the radius of a cylinder with a height of 10 cm and a curved surface area of 360π cm²?

6 What is the height of a cylinder with a diameter of 12 cm and a volume of 108π cm³?

7 A cylinder of height 20 cm has a curved surface area of 200 cm². Calculate the volume of this cylinder.

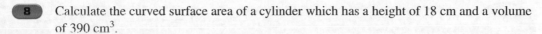

8 Calculate the curved surface area of a cylinder which has a height of 18 cm and a volume of 390 cm³.

9 A cylinder has the same height and radius. The total surface area is 100π. Calculate the volume. Give your answer in terms of π.

AU 10 A square of paper of side 10 cm is bent round to make a cylindrical shape by putting two edges together.
What is the volume of the cylinder?

FM 11 A cylindrical food can must have a volume of at least 400 cm³ in order to hold the correct amount.
The diameter of the can has to be 7 cm.
What is the smallest possible height of the can?

PS 12 Metal cylinders are made by bending rectangular sheets of metal measuring 15 cm long by 6 cm wide until the sides meet.
How many cylinders can be made from a sheet of metal that is 2 m long and 1 m wide?

Functional Maths Activity

Packaging sweets

A sweet manufacturer wants a new package for an assortment of sweets.

The package must have a volume of 1000 cm³ in order to hold the sweets.

The chosen design will be a prism.

The length has been specified as 20 cm.

The cross-section of the package will be one of three possibilities: a square, an equilateral triangle or a circle.

You have been asked to investigate the amount of packaging material needed for each design, because this will affect the cost of manufacture.

Calculate the surface area of each of the three designs.

Comment on how much difference there is between the three surface areas and how this could affect production costs.

This formula could be useful:

The area of an equilateral triangle of side a is $\frac{1}{4}\sqrt{3}a^2$

5 Algebra: Expressions and equations

5.1 Basic algebra

HOMEWORK 5A

1 Find the value of $4x + 3$ when   **a** $x = 3$ **b** $x = 6$ **c** $x = 11$

2 Find the value of $3k - 1$ when   **a** $k = 2$ **b** $k = 5$ **c** $k = 10$

3 Find the value of $4 + t$ when   **a** $t = 5$ **b** $t = 8$ **c** $t = 15$

4 Evaluate $14 - 3f$ when   **a** $f = 4$ **b** $f = 6$ **c** $f = 10$

5 Evaluate $\dfrac{4d - 7}{2}$ when   **a** $d = 2$ **b** $d = 5$ **c** $d = 15$

6 Find the value of $5x + 2$ when   **a** $x = -2$ **b** $x = -1$ **c** $x = 21.5$

7 Evaluate $4w - 3$ when   **a** $w = -2$ **b** $w = -3$ **c** $w = 2.5$

8 Evaluate $10 - x$ when   **a** $x = -3$ **b** $x = -6$ **c** $x = 4.6$

9 Find the value of $5t - 1$ when   **a** $t = 2.4$ **b** $t = -2.6$ **c** $t = 0.05$

10 Evaluate $11 - 3t$ when   **a** $t = 2.5$ **b** $t = -2.8$ **c** $t = 0.99$

11 Where $H = a^2 + c^2$, find H when   **a** $a = 3$ and $c = 4$ **b** $a = 5$ and $c = 12$

12 Where $K = m^2 - n^2$, find K when   **a** $m = 5$ and $n = 3$ **b** $m = -5$ and $n = -2$

13 Where $P = 100 - n^2$, find P when   **a** $n = 7$ **b** $n = 8$ **c** $n = 9$

14 Where $D = 5x - y$, find D when   **a** $x = 4$ and $y = 3$ **b** $x = 5$ and $y = -3$

15 Where $T = y(2x + 3y)$, find T when   **a** $x = 8$ and $y = 12$ **b** $x = 5$ and $y = 7$

16 Where $m = w(t^2 + w^2)$, find m when   **a** $t = 5$ and $w = 3$ **b** $t = 8$ and $w = 7$

FM 17 Two of the first recorded units of measurement were the *cubit* and the *palm*.
The cubit is the distance from the fingertip to the elbow and the palm is the distance across the hand.
A cubit is four and a half palms.
The actual length of a cubit varied throughout history, but it is now accepted to be 54 cm.
Noah's Ark is recorded as being 300 cubits long by 50 cubits wide by 30 cubits high.
What are the dimensions of the Ark in metres?

FM Functional Maths   **AU** (AO2) Assessing Understanding   **PS** (AO3) Problem Solving

AU 18 In this algebraic magic square, every row, column and diagonal should add up and simplify to $9a + 6b + 3c$.

$3a - 3b + 4c$	$2a + 8b + c$	$4a + b - 2c$
	$3a + 2b + c$	$2a - 2b + 7c$
$2a + 3b + 4c$		$3a + 7b - 2c$

a Copy and complete the magic square.

b Calculate the value of the 'magic number' if $a = 2$, $b = 3$ and $c = 4$.

AU 19 The rule for converting degrees Fahrenheit into degrees Celsius is:
$$C = \frac{5}{9}(F - 32)$$

a Use this rule to convert 68 °F into degrees Celsius.

b Which of the following is the rule for converting degrees Celsius into degrees Fahrenheit?
$$F = \frac{9}{5}(C + 32) \qquad F = \frac{5}{9}C + 32 \qquad F = \frac{9}{5}C + 32 \qquad F = \frac{9}{5}C - 32$$

FM 20 The formula for the cost of water used by a household each quarter is:
£32.40 + £0.003 per litre of water used.
A family uses 450 litres of water each day.

a How much is their total bill per quarter? (Take a quarter to be 91 days.)

b The family pay a direct debit of £45 per month towards their electricity costs. By how much will they be in credit or debit after the quarter?

21 Using $x = 17.4$, $y = 28.2$ and $z = 0.6$, work out the value of:

a $x = \frac{y}{z}$ **b** $\frac{x + y}{z}$ **c** $\frac{x}{z} + y$

22 **a** Laser printer cartridges cost £75 and print approximately 2500 pages. Approximately how many pence per page does it cost to run, taking only ink consumption into consideration?

b A printing specialist uses a laser printer of this type. He charges a fixed rate of £4.50 to set up the design and five pence for every page.
Explain why his profit on a print run of x pages is, in pounds, $4.5 + 0.02x$

c How much profit will the printing specialist make if he prints 2000 race entry forms for a running club?

HOMEWORK 5B

Expand these expressions.

1 $3(4 + m)$ **2** $6(3 + p)$ **3** $4(4 - y)$ **4** $3(6 + 7k)$

5 $4(3 - 5f)$ **6** $2(4 - 23w)$ **7** $7(g + h)$ **8** $4(2k + 4m)$

9 $6(2d - n)$ **10** $t(t + 5)$ **11** $m(m + 4)$ **12** $k(k - 2)$

13 $g(4g + 1)$ **14** $y(3y - 21)$ **15** $p(7 - 8p)$ **16** $2m(m + 5)$

17 $3t(t - 2)$ **18** $3k(5 - k)$ **19** $2g(4g + 3)$ **20** $4h(2h - 3)$

21 $2t(6 - 5t)$ **22** $4d(3d + 5e)$ **23** $3y(4y + 5k)$ **24** $6m^2(3m - p)$

25 $y(y^2 + 7)$ **26** $h(h^3 + 9)$ **27** $k(k^2 - 4)$ **28** $3t(t^2 + 3)$

29 $5h(h^3 - 2)$ **30** $4g(g^3 - 3)$ **31** $5m(2m^2 + m)$ **32** $2d(4d^2 - d^3)$

33 $4w(3w^2 + t)$ **34** $3a(5a^2 - b)$ **35** $2p(7p^3 - 8m)$ **36** $m^2(3 + 5m)$

37 $t^3(t + 3t)$ **38** $g^2(4t - 3g^2)$ **39** $2t^2(7t + m)$ **40** $3h^2(4h + 5g)$

FM 41 An approximate rule for converting degrees Fahrenheit into degrees Celsius is:
$$C = 0.5(F - 30)$$
   a Use this rule to convert 22 °F into degrees Celsius.
   b Which of the following is an approximate rule for converting degrees Celsius into degrees Fahrenheit?
   $F = 2(C + 30)$ $F = 0.5(C + 30)$ $F = 2(C + 15)$ $F = 2(C - 15)$

AU 42 Copy the diagram below and draw lines to show which algebraic expressions are equivalent. One line has been drawn for you.

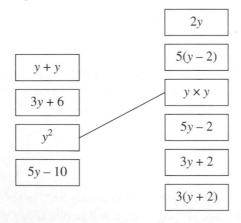

PS 43 The expansion $3(4x + 8y) = 12x + 24y$.
Write down two other expansions that give an answer of $12x + 24y$.

HOMEWORK 5C

1 Simplify these expressions.
   a $5t + 4t$ **b** $4m + 3m$ **c** $6y + y$ **d** $2d + 3d + 5d$
   e $7e - 5e$ **f** $6g - 3g$ **g** $3p - p$ **h** $5t - t$
   i $t^2 + 4t^2$ **j** $5y^2 - 2y^2$ **k** $4ab + 3ab$ **l** $5a^2d - 4a^2d$

2 Expand and simplify.

a $3(2 + t) + 4(3 + t)$ b $6(2 + 3k) + 2(5 + 3k)$ c $5(2 + 4m) + 3(1 + 4m)$

d $3(4 + y) + 5(1 + 2y)$ e $5(2 + 3f) + 3(6 - f)$ f $7(2 + 5g) + 2(3 - g)$

g $4(3 + 2h) - 2(5 + 3h)$ h $5(3g + 4) - 3(2g + 5)$ i $3(4y + 5) - 2(3y + 2)$

j $3(5t + 2) - 2(4t + 5)$ k $5(5k + 2) - 2(4k - 3)$ l $4(4e + 3) - 2(5e - 4)$

m $m(5 + p) + p(2 + m)$ n $k(4 + h) + h(5 + 2k)$ o $t(1 + 2n) + n(3 + 5t)$

p $p(5q + 1) + q(3p + 5)$ q $2h(3 + 4j) + 3j(h + 4)$ r $3y(4t + 5) + 2t(1 + 4y)$

s $t(2t + 5) + 2t(4 + t)$ t $3y(4 + 3y) + y(6y - 5)$ u $5w(3w + 2) + 4w(3 - w)$

v $4p(2p + 3) - 3p(2 - 3p)$ w $4m(m - 1) + 3m(4 - m)$ x $5d(3 - d) + d(2d - 1)$

y $5a(3b + 2a) + a(2a^2 + 3c)$ z $4y(3w + y^2) + y(3y - 4t)$

FM 3 Adult tickets for a concert cost £x and children's tickets cost £y.
At the afternoon show there were 40 adults and 160 children.
At the evening show there were 60 adults and 140 children.

a Write down an expression for the total amount of money taken on that day in terms of x and y.

b The daily expense for putting on the show is £2200. If $x = 12$ and $y = 9$, how much profit did the theatre make that day?

AU 4 Don wrote the following:
$2(3x - 1) + 5(2x + 3) = 5x - 2 + 10x + 15 = 15x - 13$
Don has made two mistakes in his working.
Explain the mistakes that Don has made.

PS 5 An internet site sells CDs. They cost £$(x + 0.75)$ each for the first five and then £$(x + 0.25)$ for any orders over five.

a Moe buys eight CDs. Which of the following expressions represents how much Moe will pay?

i $8(x + 0.75)$ ii $5(x + 0.75) + 3(x + 0.25)$

iii $3(x + 0.75) + 5(x + 0.25)$ iv $8(x + 0.25)$

b If $x = 5$, how much will Moe pay?

5.2 Factorisation

HOMEWORK 5D

Factorise the following expressions.

1 $9m + 12t$ **2** $9t + 6p$ **3** $4m + 12k$ **4** $4r + 6t$

5 $2mn + 3m$ **6** $4g^2 + 3g$ **7** $4w - 8t$ **8** $10p - 6k$

9 $12h - 10k$ **10** $4mp + 2mk$ **11** $4bc + 6bk$ **12** $8ab + 4ac$

13 $3y^2 + 4y$ **14** $5t^2 - 3t$ **15** $3d^2 - 2d$ **16** $6m^2 - 3mp$

17 $3p^2 + 9pt$ **18** $8pt + 12mp$ **19** $8ab - 6bc$ **20** $4a^2 - 8ab$

21 $8mt - 6pt$ **22** $20at^2 + 12at$ **23** $4b^2c - 10bc$ **24** $4abc + 6bed$

25 $6a^2 + 4a + 10$ **26** $12ab + 6bc + 9bd$ **27** $6t^2 + 3t + at$

28 $96mt^2 - 3mt + 69m^2t$ **29** $6ab^2 + 2ab - 4a^2b$ **30** $5pt^2 + 15pt + 5p^2t$

31 Factorise the following expressions where possible. List those which cannot factorise.

a $5m - 6t$	**b** $3m + 2mp$	**c** $t^2 - 5t$	**d** $6pt + 5ab$
e $8m^2 - 6mp$	**f** $a^2 + c$	**g** $3a^2 - 7ab$	**h** $4ab + 5cd$
i $7ab - 4b^2c$	**j** $3p^2 - 4t^2$	**k** $6m^2t + 9t^2m$	**l** $5mt + 3pn$

FM 32 An ink cartridge is priced at £9.99.

The shop has a special offer of 20% off if you buy five or more.

20% of £9.99 is £1.99.

Tom wants six cartridges. Tess wants eight cartridges.

Tom writes down the calculation $6 \times 9.99 - 6 \times 1.99$ to work out how much he must pay.

Tess writes down the calculation $8 \times (9.99 - 1.99)$ to work out how much she must pay.

Both calculations are correct.

a Who has the easier calculation and why?

b How much will each of them pay for their cartridges?

AU 33 **a** Factorise these expressions.

  i $4x + 3 + 5x - 7 - 8x$

  ii $3x - 12$

  iii $x^2 - 4x$

b What do all the answers in **a** have in common?

PS 34 To keep a class quiet a cover teacher asked them to add up all the numbers from 1 to 100 (i.e. $1 + 2 + 3 + 4 + \dots + 98 + 99 + 100$).

Two minutes later, a student said she had the correct answer.

The teacher asked the student to show the class her method.

The student wrote:

$(1 + 100) + (2 + 99) + (3 + 98) + \dots (50 + 51) = 50 \times 101$

a Explain why this gives the correct answer.

b What is the sum of all the numbers from 1 to 100?

5.3 Solving linear equations

HOMEWORK 5E

1 Solve the following equations.

a $\frac{g}{3} + 2 = 8$	**b** $\frac{m}{4} - 5 = 2$	**c** $\frac{f}{6} + 3 = 12$	**d** $\frac{h}{8} - 3 = 5$
e $\frac{2h}{3} + 3 = 7$	**f** $\frac{3t}{4} - 3 = 6$	**g** $\frac{2d}{5} + 3 = 18$	**h** $\frac{3x}{4} - 1 = 8$
i $\frac{x+5}{3} = 2$	**j** $\frac{t+12}{2} = 5$	**k** $\frac{w-3}{5} = 3$	**l** $\frac{y-9}{2} = 3$

AU 2 The solution to the equation $\frac{3x}{4} + 3 = 9$ is $x = 8$.

Make up **two** more **different** equations of the form $\frac{ax}{b} \pm c = d$, where a, b, c and d are positive whole numbers, for which the answer is also 8.

3 Solve the following equations.

a $\dfrac{2x-1}{3} = 5$ b $\dfrac{5t-4}{2} = 3$ c $\dfrac{4m+1}{5} = 5$ d $\dfrac{8p-6}{5} = 2$

e $\dfrac{5x+1}{4} = 4$ f $\dfrac{17+2t}{9} = 1$ g $\dfrac{2+4x}{3} = 4$ h $\dfrac{8-2x}{11} = 1$

FM 4 A party of eight friends went for a meal in a restaurant. The bill was £x. They received a £2 per person reduction on the bill.
They split the bill between them. Each person paid £11.25.
a Set this problem up as an equation.
b Solve the equation to work out the bill before the reduction.

HOMEWORK 5F

1 Solve the following equations. Give your answers as fractions or decimals as appropriate.

a $3(x+6) = 15$ b $6(x-5) = 30$ c $4(t+3) = 20$
d $5(4x+3) = 45$ e $3(4y-7) = 15$ f $4(5x+2) = 88$
g $3(4t+2) = 18$ h $3(4t+5) = 51$ i $4(6x+5) = 8$
j $5(3y-1) = 10$ k $5(2k+3) = 35$ l $5(2x+8) = 30$
m $3(2y-7) = 21$ n $3(2t-5) = 27$ o $8(2x-7) = 16$
p $8(3x-4) = 16$ q $4(x+7) = 8$ r $3(x-5) = -24$
s $5(t+3) = 15$ t $4(3x-13) = 8$ u $5(4t+3) = 20$
v $2(5x-3) = -16$ w $4(6y-8) = -8$ x $3(2x+7) = 9$

FM 2 A rectangular room is four metres longer than it is wide. The perimeter is 28 metres.
It cost £607.50 to carpet the room.
How much is the carpet per square metre?

AU 3 Mike has been asked to solve the equation $a(bx+c) = 60$
Mike knows that the values of a, b and c are 2, 4 and 5, but he doesn't know which is which.
He also knows that the answer is an even number.
What are the correct values of a, b and c?

PS 4 As the class are coming in for the start of a mathematics lesson, the teacher is writing some equations on the board.
So far she has written:
$5(2x+3) = 13$
$2(5x+3) = 13$
Zak says, "That's easy – both equations have the same solution, $x = 2$."
Is Zak correct? If not, then what mistake has he made? What are the correct answers?

HOMEWORK 5G

Solve each of the following equations.

1 $3x+4 = x+6$ **2** $4y+3 = 2y+5$ **3** $5a-2 = 2a+4$

4 $6t+5 = 2t+25$ **5** $8p-3 = 3p+12$ **6** $5k+4 = 2k+13$

7 $2(d+4) = d+15$ **8** $4(x-3) = 3(x+3)$ **9** $2(3y+2) = 5(2y-4)$

10 $3(2b-1) + 25 = 4(3b+1)$ **11** $3(4c+1) - 17 = 2(3c+2) - 3c$

FM 12 Olivia has some unlabelled tins of rice pudding. She needs to find out how much they weigh.

She doesn't have any weights, but she has a set of scales and some other containers with labels on them.

After some trial and improvement, she finds that five rice pudding tins and one tin of beans weighing 120 g balance with three rice pudding tins and two jars of jam that weigh 454 g each.

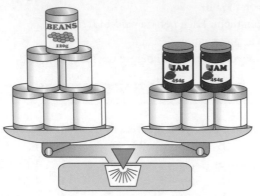

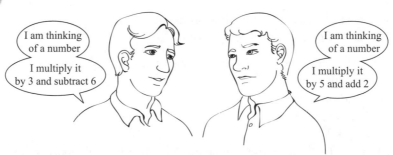

> **HINTS AND TIPS**
>
> Set up equations to work out the answers to questions 12–14.

How much does one tin of rice pudding weigh?

AU 13 The triangle shown is isosceles.

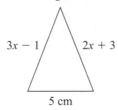

$3x - 1$ $2x + 3$

5 cm

What is the perimeter of the triangle?

PS 14

Eric and Ernie find that they both thought of the same number and both got the same final answer.

What number did they think of?

5.4 Setting up equations

HOMEWORK 5H

Set up an equation to represent each situation described below. Then solve the equation.

Do not forget to check each answer.

PS 1 A girl is Y years old. Her father is 23 years older than she is. The sum of their ages is 37. How old is the girl?

PS 2 A boy is X years old. His sister is twice as old as he is. The sum of their ages is 24. How old is the boy?

PS 3 The diagram shows a rectangle. Find x if the perimeter is 24 cm.

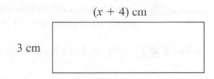

PS 4 Find the length of each side of the pentagon, if it has a perimeter of 32 cm.

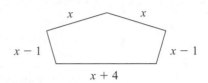

PS 5 On a bookshelf there are $2b$ crime novels, $3b - 2$ science fiction novels and $b + 7$ romance novels. Find how many of each type of book there is, if there are 65 books altogether.

PS 6 Maureen thought of a number. She multiplied it by 4 and then added 6 to get an answer of 26. What number did she start with?

PS 7 Declan also thought of a number. He took 4 from the number and then multiplied by 3 to get an answer of 24. What number did he start with?

FM 8 Books cost twice as much as magazines.
Kerry buys the same number of books and magazines and pays £22.50.
Derek buys one book and two magazines and pays £6.
How many magazines did Kerry buy?

FM 9 Sandeep's money box contains 50p coins, £1 coins and £2 coins.
In the box there are twice as many £1 coins as 50p coins and 4 more £2 coins than 50p coins. There are 44 coins in the box.
a Find how many of each coin there are in the box.
b How much money does Sandeep have in her money box?

5.5 Trial and improvement

HOMEWORK 5I

1 Find two consecutive whole numbers between which the solution to each of the following equations lies.
a $x^3 + x = 7$ **b** $x^3 + x = 55$ **c** $x^3 + x = 102$ **d** $x^3 + x = 89$

2 Find a solution to each of the following equations to 1 decimal place.
a $x^3 - x = 30$ **b** $x^3 - x = 95$ **c** $x^3 - x = 150$ **d** $x^3 - x = 333$

3 Show that $x^3 + x = 45$ has a solution between $x = 3$ and $x = 4$, and find the solution to 1 decimal place.

4 Show that $x^3 - 2x = 95$ has a solution between $x = 4$ and $x = 5$, and find the solution to 1 decimal place.

5 A rectangle has an area of 200 cm². Its length is 8 cm longer than its width. Find the dimensions of the rectangle, correct to 1 decimal place.

6 A gardener wants his rectangular lawn to be 15 m longer than the width, and the area of the lawn to be 800 m². What are the dimensions he should make his lawn? (Give your solution to 1 decimal place.)

7 A triangle has a vertical height 2 cm longer than its base length. Its area is 20 cm^2. What are the dimensions of the triangle? (Give your solution to 1 decimal place.)

8 A rectangular picture has a height 3 cm shorter than its length. Its area is 120 cm^2. What are the dimensions of the picture? (Give your solution to 1 decimal place.)

FM 9 This cuboid has a volume of 1000 cm^3.

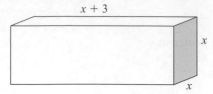

a Write down an expression for the volume.
b Use trial and improvement to find the value of x to 1 decimal place.

AU 10 Darius is using trial and improvement to find a solution to the equation $x^3 - x^2 = 25$.

The table shows his first trial.

x	$x^3 - x^2$	Comment
3	18	Too low

Continue the table to find a solution to the equation. Give your answer to 1 decimal place.

PS 11 Two numbers a and b are such that $ab = 20$ and $a - b = 5$.

Use trial and improvement to find the two numbers to one decimal place.
You can use a table like the one below. The first two lines have been done for you.

a	$b = (20 \div a)$	$a - b$	Comment
5	4	1	Too low
10	2	8	Too high

Problem-solving Activity

Throwing a ball

When a ball is thrown straight up in the air, its approximate height, h, above the ground is given by the equation $h = ut - 5t^2$, where u is the speed it is thrown at and t is the time it is in the air.

This assumes that the ball starts from the ground. If it is released from 1 m high, then the equation is $h - 1 = ut - 5t^2$

a   If $u = 16$ m/s, complete this table to show the height of the ball at every quarter second.

Time	0.25	0.5	0.75	1	1.25	......
Height	3.69	6.75	9.19	11	12.19	......

Copy the table and carry it on until the ball is back on the ground.

b   Estimate how long the ball was in the air.

c   When the ball is thrown with a speed of 16 m/s, you can work out exactly how long it was in the air.
At the start and at the end, $h = 0$.
Substitute $h = 0$ and $u = 16$ into the original equation and solve it.
What are the solutions for x?

d   Look back at your table.
Estimate the greatest height reached.

e   The greatest height reached is when the time is halfway through.
Take your answer to **c** and halve it, then substitute this value for t and $u = 16$ in the formula.
How does the answer compare with your estimate?

6.1 Pythagoras' theorem

HOMEWORK 6A

1 In each of the following triangles, find the hypotenuse, rounding off to a suitable degree of accuracy.

a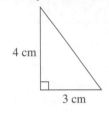

4 cm
3 cm

b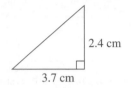

2.4 cm
3.7 cm

c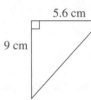

5.6 cm
9 cm

d

26 cm
24 cm

e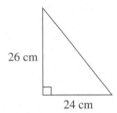

12 cm
16 cm

f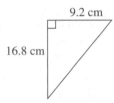

9.2 cm
16.8 cm

2 This diagram shows the cross-section of a swimming pool 50 m long. It is 3.5 m deep at the deep end. The deepest part of the pool is 10 m long. It is not drawn to scale.
  a Calculate the length of the sloping bottom of the pool AB.
  b The pool is 20 m wide. What is its volume?

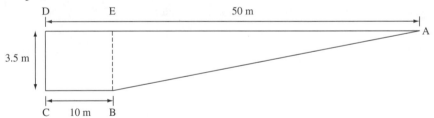

D   E   50 m

3.5 m

C   10 m   B   A

AU 3 Three of these lengths could form the sides of a right-angled triangle:
7.5 cm   10 cm   12.5 cm   15 cm
Which one would not be used? Give a reason for your answer.

FM 4 A beam of wood is needed to support a sloping roof, as shown. The beam spans a horizontal distance of 3.50 m and the difference between the bottom and the top is 1.50 m.

Beam

1.50 m

3.50 m

A builder has a beam that is 4 m long.
Is it long enough?

6.2 Finding a shorter side

HOMEWORK 6B

1 In each of the following triangles, find the length of x to a suitable degree of accuracy.

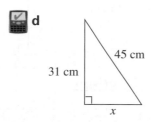
a

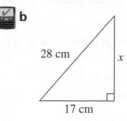
b

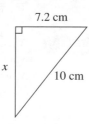
c

7.2 cm

10 cm

x

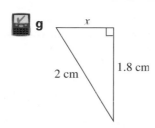
d

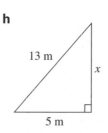
e

x

17.2 cm

19 cm

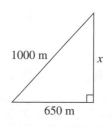
f

1000 m

x

650 m

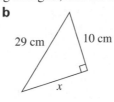
g

x

2 cm

1.8 cm

h

13 m

x

5 m

2 In each of the following triangles, find the length of x to a suitable degree of accuracy.

a

8 m

x

6 m

b

29 cm

10 cm

x

c

15 m

33 m

x

d

9.5 cm

x

8 cm

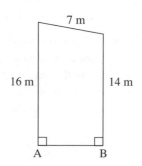
AU 3 The diagram shows the end view of the framework for a sports arena stand. Calculate the length AB.

7 m

16 m

14 m

A   B

4 Calculate the lengths of *a* and *b*.

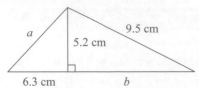

9.5 cm
a
5.2 cm
6.3 cm
b

FM 5 A ladder 3.8 m long is placed against a wall. The foot of the ladder is 1.1 m from the wall.

A window cleaner can reach windows that are 1 m above the top of her ladder. Can she reach a window that is 4 m above the ground?

6 The lengths of the three sides of a right-angled triangle are all a whole number of centimetres. The hypotenuse is 15 cm. How long are the two other sides?

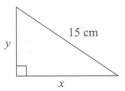

15 cm
y
x

6.3 Applying Pythagoras' theorem in real-life situations

HOMEWORK 6C

1 A ladder, 15 m long, leans against a wall. The ladder needs to reach a height of 12 m. How far should the foot of the ladder be placed from the wall?

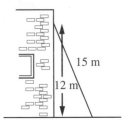

15 m
12 m

2 A rectangle is 3 m long and 1.2 m wide. How long is the diagonal?

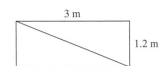

3 m
1.2 m

3 How long is the diagonal of a square with a side of 10 m?

4 A ship going from a port to a lighthouse steams 8 km east and 6 km north. How far is the lighthouse from the port?

5 At the moment, three towns, A, B and C, are joined by two roads, as in the diagram. The council wants to make a road that runs directly from A to C. How much distance will the new road save compared to the two existing roads?

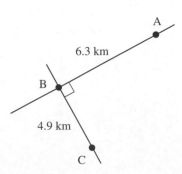

A
6.3 km
B
4.9 km
C

6 An 8 m ladder is put up against a wall.
   a How far up the wall will it reach when the foot of the ladder is 1 m away from the wall?
   b When it reaches 7 m up the wall, how far is the foot of the ladder away from the wall?

7 How long is the line that joins the two points A(1, 3) and B(2, 2)?

8 A rectangle is 4 cm long. The length of its diagonal is 5 cm. What is the area of the rectangle?

 9 Is the triangle with sides 11 cm, 60 cm and 61 cm a right-angled triangle?

10 How long is the line that joins the two points A(−3, −7) and B(4, 6)?

 AU 11 The diagram shows a voyage from position A to position B. The boat sails due east from A for 27 km to position C. The boat then changes course and sails for 30 km due south to position B. On a map, the distance between A and C is 10.8 cm.

a What is the scale of the map?

b What is the distance from A to B in kilometres?

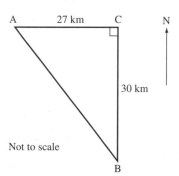

Not to scale

 AU 12 A mobile phone mast is supported by a cable that stretches from the top of the mast down to the ground. The mast is 12.5 m high and the cable is 17.8 m long.

How far from the bottom of the mast is the end of the cable that is attached to the ground?

 FM 13 A rolling pin is 45 cm long.

Will it fit inside a kitchen drawer which is internally 40 cm long and 33 cm wide?

Justify your answer.

HOMEWORK 6D

 1 Calculate the area of these isosceles triangles.

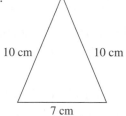

10 cm 10 cm

7 cm

5 cm

4 cm

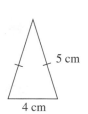

2 Calculate the area of an isosceles triangle with sides of 10 cm, 10 cm and 8 cm.

3 Calculate the area of an equilateral triangle with sides of 10 cm each.

4 **a** Calculate the area of an equilateral triangle with sides of 20 cm each.

b Explain why the answer to **4a** is not twice that of Question **3**.

5 An isosceles triangle has sides of 6 cm and 8 cm.

a Sketch the two isosceles triangles that fit this data.

b Which of the two triangles has the greater area?

6 The diagram shows an isosceles triangle of base 10 mm and side 12 mm. Calculate the area of the triangle.

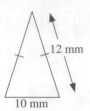

 AU 7 The diagram shows an equilateral triangle drawn inside a square with sides of 10 cm each.

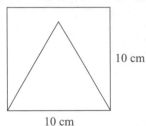

What percentage of the square is outside the triangle?

 FM 8 A picture is hanging on a string secured to two points at the side of the frame.

The string is initially 45 cm long.
When the picture is hung the string stretches as shown.
By how much does the string stretch?

6.4 Pythagoras' theorem in three dimensions

HOMEWORK 6E

1 Is the triangle with sides of 9 cm, 40 cm and 41 cm a right-angled triangle?

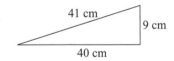

2 A box measures 6 cm by 8 cm by 10 cm.
  a Calculate the lengths of
    i AC **ii** BG **iii** BE
  b Calculate the diagonal distance BH.

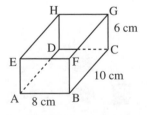

3 A garage is 5 m long, 5 m wide and 2 m high. Can a 7 m long pole be stored in it?

4 Spike, a spider, is at the corner S of the wedge shown in the diagram. Fred, a fly, is at the corner F of the same wedge.
  a Calculate the two distances Spike would have to travel to get to Fred if she used the edges of the wedge.
  b Calculate the distance Spike would have to travel across the face of the wedge to get directly to Fred.

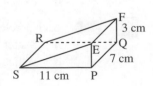

5 A corridor is 5 m wide and turns through a right angle, as in the diagram. What is the longest pole that can be carried along the corridor horizontally? If the corridor is 3 m high, what is the longest pole that can be carried along in any direction?

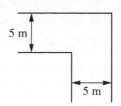

6 For the box shown on the right, find the lengths of:
  a DG
  b HA
  c DB
  d AG

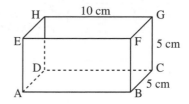

AU 7 A cube has a side of 15 cm.
Calculate the distance between two vertically opposite corners.

FM 8 A small sculpture is made from four equilateral triangles of copper sheet stuck together to make a pyramid.

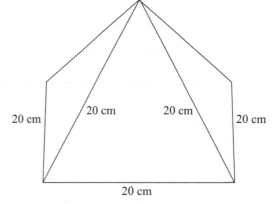

The triangles have a side of 20 cm. How high is the pyramid?

9 The diagram shows a square-based pyramid with base length 7 cm and sloping edges 12 cm. M is the mid-point of the side AB, X is the mid-point of the base, and E is directly above X.
  a Calculate the length of the diagonal AC.
  b Calculate EX, the height of the pyramid.
  c Using triangle ABE, calculate the length EM.

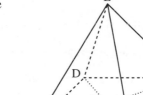

10 Use the answer to Question **1** to find the length of the diagonal AB of the cuboid 9 cm by 9 cm by 40 cm.

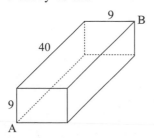

Functional Maths Activity

Access ramps

Building regulations in the UK state how steep wheelchair ramps used to access buildings are allowed to be.

Going	Maximum gradient	Maximum rise
10 m	1:20	500 mm
5 m	1:15	333 mm
2 m	1:12	166 mm

Below are some definitions of the jargon used. The diagram that follows illustrates how they are used in practice.

Going: The horizontal length.

Gradient: The tangent of the angle the ramp makes with the horizontal. Here it is written as a ratio.

Rise: The change in height from one end of the ramp to the other.

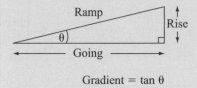

Gradient = tan θ

You can add to the table, using different lengths for the going. Try adding the maximum gradient and maximum rise for goings of 4 m and 9 m, following the number patterns in the table.

A builder has asked you to explain a few things that are puzzling him.

a Is there a connection between the numbers in the three columns?

b What is the difference between the maximum angles for a 2 m going and a 10 m going?

c He has been asked to install an access ramp to an old building. The rise required is 400 mm. In order not to exceed the available space, he wants to build a ramp with a 7 m going. Is that permitted by the regulations? Explain how you would decide.

7 Geometry: Angles and constructions

7.1 Special triangles and quadrilaterals

HOMEWORK 7A

1 For each of these shapes, calculate the value of the lettered angles.

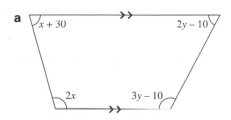

a 118° b   a   72°

b c   d   122°   e

c 161°   f   g   23°

2 Calculate the values of x and y in each of these shapes.

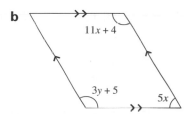

a $x + 30$ $2y - 10$ $2x$ $3y - 10$

b $11x + 4$ $3y + 5$ $5x$

PS 3 Find the value of x in each of these quadrilaterals with the following angles and state the type of quadrilateral it is.

  a $x + 10°, x + 30°, x - 30°, x - 50°$ **b** $x°, x - 10°, 3x - 15°, 3x - 15°$

4 **a** What do the interior angles of a quadrilateral add up to?
   b Use the fact that the angles of a triangle add up to 180°, to prove that the sum of the interior angles of any quadrilateral is 360°.

FM 5 The diagram shows the side wall of a barn.
The architect says that angle D must not be more than twice as big as angle A.
What is the largest possible size of angle D?

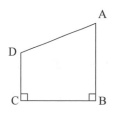

AU 6 The diagram shows a parallelogram ABCD.
AC is a diagonal.
Show that angle x is a right angle.
Give reasons for your answer.

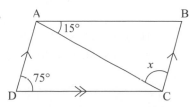

AU **7**   Give two reasons to explain why the trapezium is different from the parallelogram.

 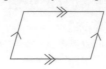

7.2 Angles in polygons

HOMEWORK 7B

1   Calculate the sum of the interior angles of polygons with:
   a   7 sides   **b**   11 sides   **c**   20 sides   **d**   35 sides

2   Calculate the size of the interior angle of regular polygons with:
   a   15 sides   **b**   18 sides   **c**   30 sides   **d**   100 sides

3   Find the number of sides of the polygon with the interior angle sum of:
   a   1440°   **b**   2520°   **c**   6120°   **d**   6840°

4   Find the number of sides of the regular polygon with an exterior angle of:
   a   20°   **b**   30°   **c**   18°   **d**   4°

5   Find the number of sides of the regular polygon with an interior angle of:
   a   135°   **b**   165°   **c**   170°   **d**   156°

PS **6**   What is the name of the regular polygon whose interior angle is three times its exterior angle?

FM **7**   Anne measured all the interior angles in a polygon. She added them up to make 1325°, but she had missed out one angle. What is the:
   a   name of the polygon that Anne measured and   **b**   size of the missing angle?

AU **8**   This shape is made from a regular pentagon and a regular octagon.

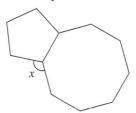

Work out the size of angle *x*.

FM **9**   Jamal is cutting metal from a rectangular sheet to make this sign.

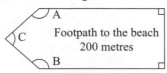

He decides it will look best if angles A and B are the same size and each of them is twice as big as angle C. How big are angles A, B and C?

PS 10 ABCDE is a regular pentagon.

Work out the size of angle ADE.
Give reasons for your answer.

 AU 11 Which of the following statements are true for a regular hexagon?

a The size of each interior angle is 60°
b The size of each interior angle is 120°
c The size of each exterior angle is 60°
d The size of each exterior angle is 240°

7.3 Constructing triangles

HOMEWORK 7C

1 Accurately draw each of the following triangles.

a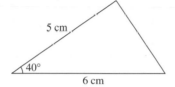

b

c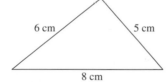

d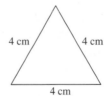

e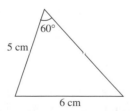

2 Draw a triangle ABC with AB = 6 cm, ∠A = 60° and ∠B = 50°.

 3

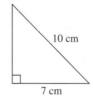

Explain why you can or cannot draw this triangle accurately.

D

4 **a** Accurately draw the shape on the right.

  b What is the name of the shape you have drawn?

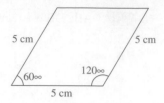

AU 5 Shehab says, "As long as I know two sides of a triangle and the angle between them then I can draw it."

Is Shehab correct?

If not, explain why not.

C

PS 6 You are asked to construct a triangle with sides 9 cm, 10 cm and an angle of 60°.

Sketch all the possible triangles that you could construct from this description.

7.4 Bisectors

HOMEWORK 7D

C

1 Draw a line 8 cm long. Bisect it with a pair of compasses. Check your accuracy by seeing if each half is 4 cm.

2 **a** Draw any triangle.

  b On each side construct the line bisector. All your line bisectors should intersect at the same point.

  c See if you can use this point as the centre of a circle that fits perfectly inside the triangle.

3 **a** Draw a circle with a radius of about 4 cm.

  b Draw a quadrilateral such that the vertices (corners) of the quadrilateral are on the circumference of the circle.

  c Bisect two of the sides of the quadrilateral. Your bisectors should meet at the centre of the circle.

4 **a** Draw any angle.

  b Construct the angle bisector.

  c Check how accurate you have been by measuring each half.

5 The diagram shows a park with two ice-cream sellers A and B. People always go to the ice-cream seller nearest to them. Shade the region of the park from which people go to ice-cream seller B.

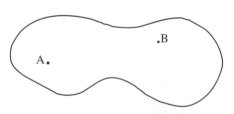

PS 6 Using a straight edge and a pair of compasses only, construct:

  a an angle of 15 degrees

  b an angle of 75 degrees

AU 7 If I construct all the angle bisectors in a triangle, they will meet at a point.

Explain why I can draw a circle with this as the centre, and this circle will just touch each side of the triangle.

7.5 Defining a locus

HOMEWORK 7E

1 A is a fixed point. Sketch the locus of the point P when AP > 3 cm and AP < 6 cm.

2 A and B are two fixed points 4 cm apart. Sketch the locus of the point P for the following situations:
   a AP < BP   **b** P is always within 3 cm of A and within 2 cm of B.

3 A fly is tethered by a length of spider's web that is 1 m long. Describe the locus that the fly can still move around in.

4 ABC is an equilateral triangle of side 4 cm. In each of the following loci, the point P moves only inside the triangle. Sketch the locus in each case.
   a AP = BP   **b** AP < BP
   c CP < 2 cm   **d** CP > 3 cm and BP > 3 cm

5 A wheel rolls around the inside of a square. Sketch the locus of the centre of the wheel.

6 The same wheel rolls around the outside of the square. Sketch the locus of the centre of the wheel.

7 Two ships A and B, which are 7 km apart, both hear a distress signal from a fishing boat. The fishing boat is less than 4 km from ship A and is less than 4.5 km from ship B.
A helicopter pilot sees that the fishing boat is nearer to ship A than to ship B. Use accurate construction to show the region which contains the fishing boat. Shade this region.

PS 8 On a piece of plain paper, mark three points A, B and C, about 5 to 7 cm away from each other.
Find the locus of point P where:
   a P is always closer to a point A than a point B
   b P is always equal distances from points B and C

AU 9 Sketch the locus of a point on the rim of a bicycle wheel as it makes three revolutions along a flat road.

7.6 Loci problems

HOMEWORK 7F

For Questions 1 to 3, you should start by sketching the picture given in each question on a 6 × 6 grid, each square of which is 1 cm by 1 cm. The scale for each question is given.

1 A goat is tethered by a rope, 10 m long, and a stake that is 2 m from each side of a field. What is the locus of the area that the goat can graze? Use a scale of 1 cm : 2 m.

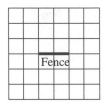

2 A cow is tethered to a rail at the top of a fence 4 m long. The rope is 4 m long. Sketch the area that the cow can graze. Use a scale of 1 cm : 2 m.

3 A horse is tethered to a corner of a shed, 3 m by 1 m. The rope is 4 m long. Sketch the area that the horse can graze. Use a scale of 1 cm : 1 m.

For Questions 4 to 7, you should use a copy of the map on page 60. For each question, trace the map and mark on those points that are relevant to that question.

4 A radio station broadcasts from Birmingham with a range that is just far enough to reach York. Another radio station broadcasts from Glasgow with a range that is just far enough to reach Newcastle.
  a Sketch the area to which each station can broadcast.
  b Will the Birmingham station broadcast as far as Norwich?
  c Will the two stations interfere with each other?

5 An air traffic control centre is to be built in Newcastle. If it has a range of 200 km, will it cover all the area of Britain north of Sheffield and south of Glasgow?

FM 6 A radio transmitter is to be built so that it is the same distance from Exeter, Norwich and Newcastle.
  a Draw the perpendicular bisectors of the lines joining these three places to find where it is to be built.
  b Birmingham has so many radio stations that it cannot have another one within 50 km. Can the transmitter be built?

PS 7 Three radio stations receive a distress call from a boat in the North Sea.
The station at Norwich can tell from the strength of the signal that the boat is within
150 km of the station. The station at Sheffield can tell that the boat is between 100 and
150 km from Sheffield.
If these two reports are correct, then how far away from the helicopter station at
Newcastle might the boat be?

AU 8 The locus of a point is described as:
5 cm away from point A
Equidistant from both points A and B
Which of the following could be true?
   a   The locus is an arc
   b   The locus is just two points
   c   The locus is a straight line
   d   The locus is none of these

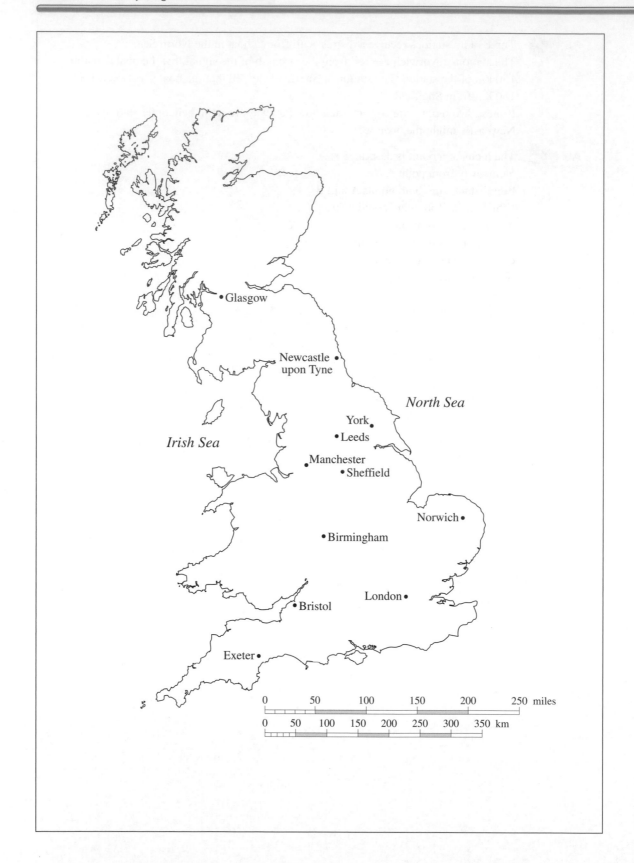

Problem-solving Activity

The nine-point circle

You will have already constructed the circumcircle and circumscribed circle of a triangle. This construction is the nine-point circle of a triangle.

Step 1
Draw a triangle ABC.

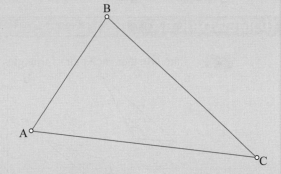

Step 2
Construct the mid-points of the three sides and call these L for AB, M for BC and N for AC.

Step 3
Construct the perpendiculars to the opposite sides from the vertices A, B and C. Call the point where they intersect O. Label the feet of the perpendiculars D, E and F on AB, BC and AC respectively.

Step 4
Construct the midpoints of AO, BO and CO. Label these X, Y and Z respectively.

Step 5
Bisect the line segments LM, LN and MN. Call the point where they intersect P.

Step 6
Can you think of a way to check all nine points you have drawn?

8.1 Congruent triangles

HOMEWORK 8A

1 State whether each pair of triangles below is congruent, giving reasons if they are.

a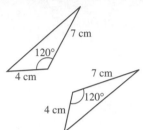

b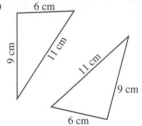

c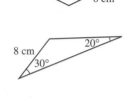

2 Draw a square ABCD. Draw in the diagonals AC and BD. Which triangles are congruent to each other?

3 Draw a kite EFGH. Draw in the diagonals EG and FH. Which triangles are congruent to each other?

4 Draw a rhombus ABCD. Draw in the diagonals AC and BD. Which triangles are congruent to each other?

5 Draw an equilateral triangle ABC. Draw the lines from each vertex to the mid-point of the opposite side. These three lines should all cross at the same point T inside the triangle. Which triangles are congruent to each other?

PS 6 In the diagram, AB and CD are parallel with AB = CD.
The lines AC and BD intersect at X.

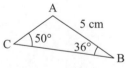

Prove that triangle ABX and triangle CDX are congruent.

AU 7 Helen says that these two triangles are congruent because the three angles are the same.

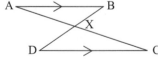

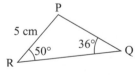

Explain why she is wrong.

FM Functional Maths **AU** (AO2) Assessing Understanding **PS** (AO3) Problem Solving

8.2 Translations

HOMEWORK 8B

1 Describe these translations with vectors.
   i A to B   **ii** A to C   **iii** A to D   **iv** B to A   **v** B to C   **vi** B to D

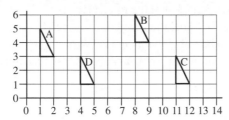

2 **a** On a grid showing values of x and y from 0 to 10, draw the triangle with coordinates A(4, 4), B(5, 7) and C(6, 5).

   b Draw the image of ABC after a translation with vector $\binom{3}{2}$. Label this P.

   c Draw the image of ABC after a translation with vector $\binom{4}{-3}$. Label this Q.

   d Draw the image of ABC after a translation with vector $\binom{-4}{3}$. Label this R.

   e Draw the image of ABC after a translation with vector $\binom{-3}{2}$. Label this S.

3 Using your diagram from Question **2**, describe the translation that will move
   a P to Q   **b** Q to R   **c** R to S   **d** S to P
   e R to P   **f** S to Q   **g** R to Q   **h** P to S

FM 4 A group of hikers walk between three points A, B and C using direction vectors with distances in kilometres.

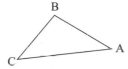

The direction vector from A to B is $\binom{-4}{3}$ and the direction vector from B to C is $\binom{-2}{-5}$.
On centimetre squared paper, draw a diagram to show the walk, using a scale of 1 cm represents 1 km.
Work out the direction vector from C to A.

PS 5 Write down a series of translations which will take you from the Start/finish, around the shaded square without touching it, and back to the Start/finish. Make as few translations as possible.

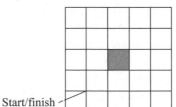

Start/finish

AU 6 Joel says that if the translation from a point X to a point Y is described by the vector $\binom{-3}{2}$, then the translation from the point Y to the point X is described by the vector $\binom{2}{-3}$.
Is Joel correct? Explain how you decide.

8.3 Reflections

HOMEWORK 8C

1 **a** Draw a pair of axes with the *x*-axis from –5 to 5 and the *y*-axis from –5 to 5.
b Draw the triangle with co-ordinates A(1, 1), B(5, 5), C(3, 4).
c Reflect triangle ABC in the *x*-axis. Label the image P.
d Reflect triangle P in the *y*-axis. Label the image Q.
e Reflect triangle Q in the *x*-axis. Label the image R.
f Describe the reflection that will transform triangle ABC onto triangle R.

AU 2 Copy this diagram onto squared paper.

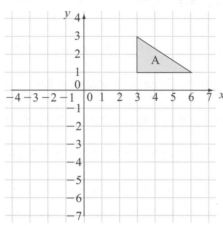

a Reflect triangle A in the line $x = 1$
Label the image B.
b Reflect triangle B in the line $y = -2$
Label the image C.

FM 3 A designer is making a logo for a company.
She starts with a kite ABCD.
She then reflects the kite in the line BD on top of the
original kite to obtain the logo.
Draw any kite on squared paper to obtain a logo by
following the designer's method.

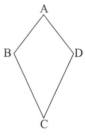

PS 4 A point X has coordinates (a, b).
Point X is reflected in the line $x = 3$
Find the coordinates of the image of point X.

5 **a** Draw a pair of axes, *x*-axis from –5 to 5, *y*-axis from –5 to 5.
b Draw the triangle with co-ordinates A(2, 2), B(3, 4), C(2, 4).
c Reflect the triangle ABC in the line $y = x$. Label the image P.
d Reflect the triangle P in the line $y = -x$. Label the image Q.
e Reflect triangle Q in the line $y = x$. Label the image R.
f Describe the reflection that will move triangle ABC to triangle R.

8.4 Rotations

HOMEWORK 8D

1 Copy this T-shape on squared paper.

a Rotate the shape 90° clockwise about the origin 0. Label the image P.

b Rotate the shape 180° clockwise about the origin 0. Label the image Q.

c Rotate the shape 90° anticlockwise about the origin 0. Label the image R.

d What rotation takes R back to the original shape?

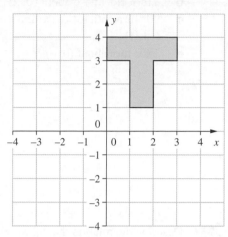

2 Copy this square ABCD on squared paper.

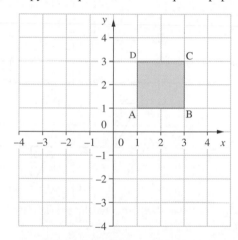

a Write down the coordinates of the vertices of the square ABCD.

b Rotate the square ABCD through 90° clockwise about the origin 0. Label the image S. Write down the coordinates of the vertices of the square S.

c Rotate the square ABCD through 180° clockwise about the origin 0. Label the image T. Write down the coordinates of the vertices of the square T.

d Rotate the square ABCD through 90° anticlockwise about the origin 0. Label the image U. Write down the coordinates of the vertices of the square U.

e What do you notice about the coordinates of the four squares?

3 A designer is making a logo for a company.
She starts with a parallelogram ABCD.
She then rotates the parallelogram 90° clockwise about the point of intersection of the two diagonals on top of the original parallelogram to obtain the logo.
Draw any parallelogram on squared paper to obtain a logo by following the designer's method.

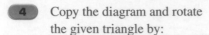

4 Copy the diagram and rotate the given triangle by:

    a $\frac{1}{4}$ turn clockwise about (0, 0)

    b $\frac{1}{2}$ turn clockwise about (0, 2)

    c 90° turn anticlockwise about (−1, 1)

    d 180° turn about (0, 0).

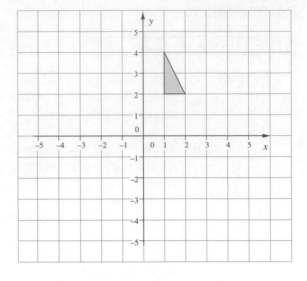

5 Describe the rotation that takes the shaded triangle to:

    a triangle A

    b triangle B

    c triangle C

    d triangle D.

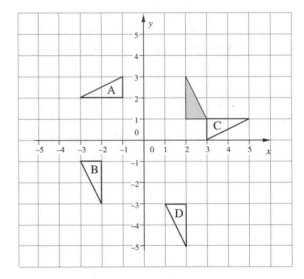

PS 6 A point P has coordinates (a, b).

    a The point P is rotated 90° clockwise about (0, 0) to give a point Q. What are the coordinates of Q?

    b The point P is rotated 180° clockwise about (0, 0) to give a point R. What are the coordinates of R?

    c The point P is rotated 90° anticlockwise about (0, 0) to give a point S. What are the coordinates of S?

AU 7 Triangle A is drawn on a grid.
Triangle A is rotated to form a new triangle B.
The coordinates of B are (3, –1), (1, –4)
and (3, –4).
Describe fully the rotation that maps
triangle A onto triangle B.

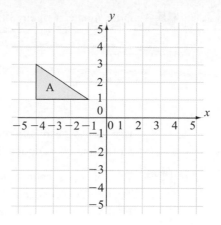

8.5 Enlargements

HOMEWORK 8E

1 Copy each figure below with its centre of enlargement, leaving plenty of space for the enlargement. Then enlarge them by the given scale factor, using the ray method.

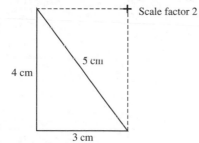

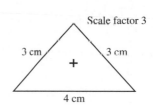

2 Copy each of these diagrams on squared paper and enlarge it by scale factor 2 using the origin as the centre of enlargement.

a

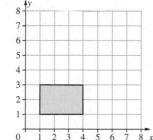

b

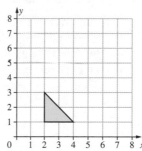

c

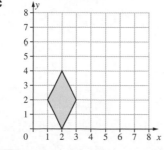

d

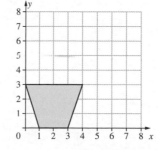

3   Draw a letter T of any size. Now draw another letter T twice the size, as in the diagram.

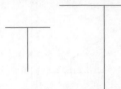

Use the ray method to find the centre of enlargement.
Draw the rays as dotted lines to create a logo design.

AU 4   Triangle A has coordinates (2, 2), (6, 2) and (6, 4).
Triangle A is enlarged by a scale factor of $\frac{1}{2}$ about the origin to give triangle B.
Find the coordinates of triangle B.

AU 5   Triangle B is an enlargement of triangle A.

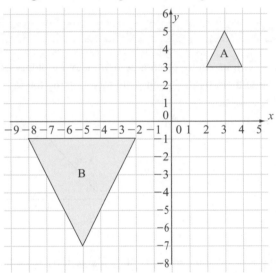

Which of the following describes the enlargement?
a   an enlargement of scale factor –2 about (0, 0).
b   an enlargement of scale factor –3 about (0, 0).
c   an enlargement of scale factor –3 about (1, 2).
d   an enlargement of scale factor $-\frac{1}{3}$ about (1, 2).
     Show how you decide.

8.6 Combined transformations

1 Describe fully the transformation that will move
 a T_1 to T_2 **b** T_1 to T_6 **c** T_2 to T_3 **d** T_6 to T_2
 e T_6 to T_5 **f** T_5 to T_4 **g** T_1 to T_5

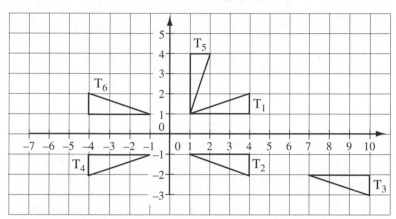

2 **a** Plot a triangle T with vertices (1, 1), (3, 1), (3, 4).
 b Reflect triangle T in the x-axis and label the image T_b.
 c Rotate triangle T_b 90° clockwise about the origin and label the image T_c.
 d Reflect triangle T_c in the x-axis and label the image T_d.
 e Describe fully the transformation that will move triangle T_d back to triangle T.

PS 3 **a** The point P(2, 5) is reflected in the x-axis, then rotated by 90° clockwise about the origin.
 What are the coordinates of the image of P?
 b The point Q(a, b) is reflected in the x-axis, then rotated by 90° clockwise about the origin. What are the coordinates of the image of Q?

PS 4 **a** The point R(4, 3) is reflected in the line $y = -x$, then reflected in the x-axis. What are the coordinates of the image of R?
 b The point S(a, b) is reflected in the line $y = -x$, then reflected in the x-axis. What are the coordinates of the image of S?

AU 5   Copy the diagram onto squared paper.

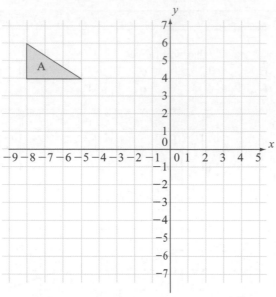

a   Triangle A is translated by the vector $\binom{9}{-3}$ to give triangle B.
Triangle B is then enlarged by a scale factor –2 about the origin to give triangle C.
Draw triangles B and C on the diagram.

b   Describe fully the single transformation that maps triangle C onto triangle A.

Functional Maths Activity

Transformations in the sorting office

When letters are taken to a sorting office they are checked to see if there is a stamp in the top right-hand corner.

Each letter is put through a checking machine and the top right-hand corner is scanned. Of course, the stamp will be detected only if the letter is the right way round.

If no stamp is detected, the letter is rotated automatically and put through the machine again.

Two different rotations are used:

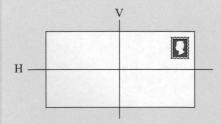

- 180° rotation about the horizontal line H
- 180° rotation about the vertical line V

These rotations will leave the letter in the same orientation, but with the stamp in the other corner.

Functional Maths Activity (continued)

Here is the procedure used:

- Scan the letter
- If no stamp is detected, rotate about H and scan again
- If no stamp is detected, rotate about V and scan again
- If no stamp is detected, rotate about H and scan again
- If no stamp is detected then the letter is rejected as unstamped.

1   Show that if this procedure is followed then any letter that is correctly stamped will be detected, whichever way the letter is initially fed into the machine.

2   If someone accidentally put the stamp on the top left-hand corner of the letter, would the machine detect it?

3   Suppose that a new regulation requires all letters to be square. Explain why there are now eight corners to be checked for a stamp.

4   With a square letter you could use rotations about either of the diagonals as well as the horizontal and vertical lines. Show how it is still possible to use just two rotations (repeated if necessary) to check all eight corners.

9 Statistics: Data handling

9.1 Averages

1 a For each set of data find the mode, the median and the mean.

    i   6, 4, 5, 6, 2, 3, 2, 4, 5, 6, 1

    ii 14, 15, 15, 16, 15, 15, 14, 16, 15, 16, 15

    iii 31, 34, 33, 32, 46, 29, 30, 32, 31, 32, 33

b For each set of data decide which average is the best one to use and give a reason.

2 A supermarket sells oranges in bags of ten.

The weights of each orange in a selected bag were as follows:

    134 g, 135 g, 142 g, 153 g, 156 g, 132 g, 135 g, 140 g, 148 g, 155 g

a Find the mode, the median and the mean for the weight of the oranges.

b The supermarket wanted to state the average weight on each bag they sold. Which of the three averages would you advise the supermarket to use? Explain why.

3 The weights, in kilograms, of players in a school football team are as follows:

    68, 72, 74, 68, 71, 78, 53, 67, 72, 77, 70

a Find the median weight of the team.

b Find the mean weight of the team.

c Which average is the better one to use? Explain why.

4 Jez is a member of a local quiz team and, in the last eight games, his total points were:

    62, 58, 24, 47, 64, 52, 60, 65

a Find the median for the number of points he scored over the eight weeks.

b Find the mean for the number of points he scored over the eight weeks.

c The team captain wanted to know the average for each member of the team. Which average would Jez use? Give a reason for your answer.

FM 5 Three dancers were hoping to be chosen to represent their school in a competition.
They had all been involved in previous competitions.
The table below shows their scores in recent contests.

Kathy	8, 5, 6, 5, 7, 4, 5
Connie	8, 2, 7, 9, 2
Evie	8, 1, 8, 2, 3

The teachers said they would be chosen by their best average score.
Which average would each dancer prefer to be chosen by?

PS 6 a Find 3 numbers that have **all** the properties below:

  • a range of 3

  • a mean of 3

b Find 3 numbers that have **all** the properties below:

  • a range of 3

  • a median of 3

  • a mean of 3

FM Functional Maths   **AU** (AO2) Assessing Understanding   **PS** (AO3) Problem Solving

AU 7 A class of students were taking a test.
When asked, "What is the average score for the test?" the teacher said 32 but a student said 28.
They were both correct. Explain how this could be.

9.2 Frequency tables

1 Find **i** the mode, **ii** the median and **iii** the mean from each frequency table below.

   a A survey of the collar sizes of all the male staff in a school gave these results.

Collar size	12	13	14	15	16	17	18
Number of staff	1	3	12	21	22	8	1

   b A survey of the number of TVs in pupils' homes gave these results.

Number of TVs	1	2	3	4	5	6	7
Frequency	12	17	30	71	96	74	25

2 A survey of the number of pets in each family of a school gave these results.

Number of pets	0	1	2	3	4	5
Frequency	28	114	108	16	15	8

   a Each child at the school is shown in the data. How many children are at the school?
   b Calculate the median number of pets in a family.
   c How many families have less than the median number of pets?
   d Calculate the mean number of pets in a family. Give your answer to 1 dp.

AU 3 Twinkle travelled to Manchester on many days throughout the year.
The table shows how many days she travelled in each week.

Days	0	1	2	3	4	5
Frequency (no. of weeks)	17	2	4	13	15	1

   Explain how you would find the median number of days that Twinkle travelled in a week to Manchester.

4 A survey of the number of television sets in each family home in one school year gave these results:

Number of TVs	0	1	2	3	4	5
Frequency	1	5	36	86	72	56

   a How many students are in that school year?
   b Calculate the mean number of TVs in a home for this school year.
   c How many homes have this mean number of TVs (if you round the mean to the nearest whole number)?
   d What percentage of homes could consider themselves average from this survey?

PS 5 A coffee stain removed four numbers (in two columns) from the following frequency table of eggs laid by 20 hens one day.

Eggs	0	1	2		5
Frequency	2	3	4		1

The mean number of eggs laid was 2.5.
What could the missing four numbers be?

9.3 Grouped data

HOMEWORK 9C

1 For each table of values given below, find:
　　i　the modal group　　**ii**　an estimate for the mean.

a

Score	0 – 20	21 – 40	41 – 60	61 – 80	81 – 100
Frequency	9	13	21	34	17

b

Cost (£)	0.00 – 10.00	10.01 – 20.00	20.01 – 30.00	30.01 – 40.00	40.01 – 60.00
Frequency	9	17	27	21	14

FM 2 A hospital has to report the average waiting time for patients in the Accident and Emergency department. A survey was made to see how long casualty patients had to wait before seeing a doctor.
The following table summarises the results for one shift.

Time (minutes)	0 – 10	11 – 20	21 – 30	31 – 40	41 – 50	51 – 60	61 – 70
Frequency	1	12	24	15	13	9	5

a How many patients were seen by a doctor in the survey of this shift?
b Estimate the mean waiting time taken per patient.
c Which average would the hospital use for the average waiting time?
d What percentage of patients did the doctors see within the hour?

FM 3 A shoe shop undertook a survey to see what size men's shoes were being sold in one month. The following table summarises the results.

Shoe size	3 – 4	5 – 6	7 – 8	9 – 10	11 – 12
Frequency	2	4	21	55	32

a How many pairs of men's shoes were sold during this month?
b Estimate the mean men's shoe size sold.
c Which of the averages is of most use to the shop's manager?
d What percentage of men's shoes sold were smaller than size 7?

PS 4 The table below shows the total weight of fish caught by the anglers in a fishing competition.

Weight (Kg)	0 < w ≤ 5	5 < w ≤ 10	10 < w ≤ 15	15 < w ≤ 20	20 < w ≤ 25
Frequency	4	15	10	8	3

Helen noticed that two numbers were in the wrong part of the table and that this made a difference of 0.625 to the arithmetic mean.
Which two numbers were the wrong way round?

AU 5 The profit made each week by a tea shop is shown in the table below.

Profit	£0 – £200	£201 – £400	£401 – £600	£601 – £800
Frequency	15	26	8	3

Explain how you would estimate the mean profit made each week.

9.4 Frequency diagrams

HOMEWORK 9D

1 The table shows the time taken by 60 people to travel to work.

Time in minutes	10 or less	Between 10 and 30	30 or more
Frequency	8	19	33

Draw a pie chart to illustrate the data.

2 The table shows the number of GCSE passes that 180 students obtained.

GCSE passes	9 or more	7 or 8	5 or 6	4 or less
Frequency	20	100	50	10

Draw a suitable chart to illustrate the data.

3 Tom is doing a statistics project on the use of computers. He decides to do a survey to find out the main use of computers by 36 of his school friends. His results are shown in the table.

Main use	e-mail	Internet	Word processing	Games
Frequency	5	13	3	15

a Draw a pie chart to illustrate his data.

b What conclusions can you draw from his data?

c Give reasons why Tom's data is not really suitable for his project.

4 In a survey, a TV researcher asks 120 people at a leisure centre to name their favourite type of television programme. The results are shown in the table.

Type of programme	Comedy	Drama	Films	Soaps	Sport
Frequency	18	11	21	26	44

a Draw a pie chart to illustrate the data.

b Do you think the sample chosen by the researcher is representative of the population? Give a reason for your answer.

FM 5 Marion is writing an article on health for a magazine. She asked a sample of people the question: 'When planning your diet, do you consider your health?' The pie chart shows the results of her question.

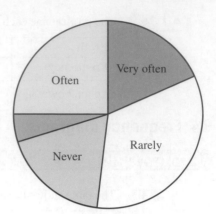

a   What percentage of the sample responded 'often'.

b   What response was given by about a third of the sample?

c   Can you tell how many people there were in the sample? Give a reason for your answer.

d   What other questions could Marion ask?

PS 6 A nationwide survey was taken on where people thought the friendliest people were in England. What is the probability that a person picked at random from this survey answered 'East'?

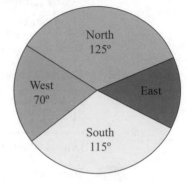

AU 7 You are asked to draw a pie chart representing the different breakfasts that students have in a morning.

What data would you need to obtain in order to do this?

HOMEWORK 9E

1 a   The table shows the ages of 300 people at the cinema.

Age, x years	$0 \leqslant x < 10$	$10 \leqslant x < 20$	$20 \leqslant x < 30$	$30 \leqslant x < 40$	$40 \leqslant x < 50$
Frequency	25	85	115	45	30

Draw a histogram to show the data.

b   At another film show this was the distribution of ages.

Age, x years	$20 \leqslant x < 30$	$30 \leqslant x < 40$	$40 \leqslant x < 50$	$50 \leqslant x < 60$	$60 \leqslant x < 70$
Frequency	35	120	130	50	15

Draw a histogram to show this data.

c   Comment on the differences between the distributions.

2 The table shows the times taken by 50 children to complete a multiplication square.

Time, s seconds	$10 \leqslant s < 20$	$20 \leqslant s < 30$	$30 \leqslant s < 40$	$40 \leqslant s < 50$	$50 \leqslant s < 60$
Frequency	3	9	28	6	4

a   Draw a frequency polygon for this data.

b   Calculate an estimate of the mean of the data.

3 The waiting times for customers at a supermarket checkout are shown in the table.
  a Draw a histogram of these waiting times.
  b Estimate the mean waiting time.

Waiting time (minutes)	Frequency
$0 \leqslant x < 2$	15
$2 \leqslant x < 4$	7
$4 \leqslant x < 6$	12
$6 \leqslant x < 8$	15
$8 \leqslant x < 10$	12

FM 4 After a mental arithmetic test, all the results were collated for girls and boys separately as shown in this table. The school wants to be able to look at the differences in the results in a clear diagram and use the diagram to estimate the mean scores.

Number correct, N	$0 \leqslant N \leqslant 4$	$5 \leqslant N \leqslant 8$	$9 \leqslant N \leqslant 12$	$13 \leqslant N \leqslant 16$	$17 \leqslant N \leqslant 20$
Boys	5	9	23	28	17
Girls	6	10	19	25	22

  a Draw frequency polygons to illustrate the differences between the boys' scores and the girls' scores.
  b Estimate the mean score for boys and girls scores separately, then comment on your results.

PS 5 The frequency polygon shows the length of time that students spent playing sport in one weekend.
Calculate an estimate of the mean time spent playing sport by the students.

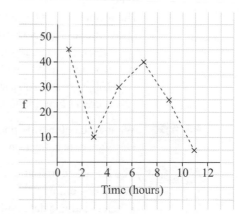

AU 6 The frequency polygon shows the times that a number of people waited at the bus stop before their bus came one morning.
Dan said, 'Most people spent 5 minutes waiting.'
Explain why this is incorrect.

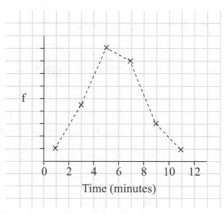

9.5 Surveys

HOMEWORK 9F

1 'People like the DVD rental shop to be open 24 hours a day.'
  a To see whether this statement is true, design a data collection sheet which will allow you to capture data while standing outside a video hire centre.
  b Does it matter at which time you collect your data?

2 The youth club wanted to know which types of activities it should plan, e.g. craft, swimming, squash, walking, disco etc.
  a Design a data collection sheet which you could use to ask the pupils in your school which activities they would want in a youth club.
  b Invent the first 30 entries on the chart.

3 What types of film do your age group watch at the cinema the most? Is it comedy, romance, sci-fi, action, suspense or something else?
  a Design a data collection sheet to be used in a survey of your age group.
  b Invent the first 30 entries on your sheet.

FM 4 Rodrigo, who works for a marketing company, wants to find out who goes to rock concerts. Identifying the target market enables him to make his marketing campaign more successful.
He decides to investigate the hypothesis:
'Boys are less likely to go to rock concerts than girls'.
  a Design a data capture form that Rodrigo could use to help him do this.
  b Rodrigo records information from a sample of 300 boys and 220 girls. He finds that 190 boys and 160 girls go to rock concerts. Based on this sample, is the hypothesis correct? Explain your answer.

PS 5 What kinds of mobile phones do your classmates own?
Design a data collection sheet to help you find this out.

AU 6 You are asked to find out what kinds of goods the parents of the students at your school like to buy online.
When creating a data collection form, what two things must you include?

9.6 Questionnaires

HOMEWORK 9G

1 Design a questionnaire to test the following statement.
'Young people aged 16 and under will not tell their parents when they have been given detention at school, but the over 16s will always let their parents know.'

2 'Boys will use the Internet almost every day but girls will only use it about once a week.'
Design a questionnaire to test this statement.

3 Design a questionnaire to test the following hypothesis.
'When you are in your twenties, you watch less TV than any other age group.'

4 While on holiday in Wales, I noticed that in the supermarkets there were a lot more women than men, and the only men I did see were over 65.
  a Write down a hypothesis from the above observation.
  b Design a questionnaire to test your hypothesis.

FM 5 Yohanes and Jacob are doing a survey on the type of DVDs people buy.

   a This is one question from Yohanes's survey.

Westerns are only watched by old people.
Don't you agree?
Strongly agree ☐   Agree ☐   Don't know ☐

      Give two criticisms of Yohanes's question.

   b This is a question from Jacob's survey.

How many DVDs do you buy each month?
3 or fewer ☐   4 or 5 ☐   more than 5 ☐

      Give two reasons why this is a good question.

   c Make up another good question with responses that could be added to this survey.

PS 6 Design a questionnaire to test the hypothesis,
'Only old people go to Tango lessons'.

AU 7 Kath left a football match and was given a questionnaire containing the following question:
Explain why the response section of this questionnaire is poor.

Question:	How many times do you come to this football ground?	
Response:	Never ☐	More than 5 times ☐
	More than 10 times ☐	More than 20 times ☐

9.7 The data-handling cycle

HOMEWORK 9H

Use the data-handling cycle to describe how you would test each of the following hypotheses, stating in each case whether you would use Primary or Secondary data.

1 January is the coldest month of the year

2 Girls are better than boys at estimating weights.

3 More men go to cricket matches than women.

4 The TV show *Strictly Come Dancing* is watched by more women than men.

5 The older you are the more likely you are to go ballroom dancing.

9.8 Other uses of statistics

HOMEWORK 9I

FM 1 In 2005, the cost of a litre of diesel was 84p. Using 2005 as a base year, the price index of diesel for the next five years is shown in the table.

Year	2005	2006	2007	2008	2009	2010
Index	100	104	108	111	113	118
Price	89p					

Work out the price of diesel in each subsequent year. Give your answers to one decimal place.

PS 2 A country's retail index was given as:
1990: 100
2000: 110
2010: 125
A meal out for four cost £62.50 in the year 2010.
What was the cost of the meal in the year 2000?

AU 3 The Retail Price Index measures how much the daily cost of living increases or decreases. If 2009 is given a base index number of 100, and 2010 given 102, what does this mean?

9.9 Sampling

HOMEWORK 9J

1 For a school project you have been asked to do a presentation of the timing of the school day. You decide to interview a sample of pupils. How will you choose those you wish to interview if you want your results to be reliable? Give three reasons for your decisions.

2 Comment on the reliability of the following ways of finding a sample.
 a Asking the year 11 Religious Education option class about religion.
 b Find out how many homes have microwaves by asking the first 100 pupils who walk through the school gates.
 c Find the most popular Playstation game by asking a Year 7 form which their favourite is.

3 Comment on the way the following samples have been taken. For those that are not satisfactory, suggest a better way to find a more reliable sample.
 a Bill wanted to find out what proportion of his school went to the cinema, so he obtained an alphabetical list of students and sent a questionnaire to every tenth person on the list.
 b The council wants to know about sports facilities in an area so they sent a survey team to the local shopping centre one Monday morning.
 c A political party wanted to know how much support they had in an area so they rang 500 people from the phone book in the evening.

4 Shameela made a survey of pupils in her school. The size of each year group in the school is shown on the right.

Claire took a sample of 150 pupils.

Year	Boys	Girls	Total
7	154	137	291
8	162	156	318
9	134	160	294
10	153	156	309
11	130	140	270
Total	733	749	1482

a Explain why this is a suitable size of sample.

b Draw up a table showing how many pupils of each sex and year she should ask if she wants to obtain a stratified sample.

5 A train company attempted to estimate the number of people who travel by train in a certain town. They telephoned 200 people in the town one evening and asked, 'Have you travelled by train in the last week?' 32 people said 'Yes.' The train company concluded that 16% of the town's population travel by train. Give three criticisms of this method of estimation.

FM 6 Mrs Reynolds, the deputy headteacher at Bradway School, wanted to find out how often the sixth form students in her school went out of the school for their lunch. The number of students in each sixth form year are given in the table.

	Boys	Girls
Y12	88	92
Y13	83	75

a Create a questionnaire that Mrs Reynolds could use to sample the school.

b Mrs Reynolds wanted to do a stratified sample using 50 students. How many of each group should she give the questionnaire to?

PS 7 Sandila's school had a total of 1260 students and there were 28 students in her class. One day a survey was carried out using students sampled from the whole school, and four boys and three girls in Sandila's class were chosen to take part in the survey. Estimate how many students in the whole school were involved in the sample.

AU 8 You are asked to conduct a survey at a concert where the attendance is approximately 15 000. Explain how you could create a stratified sample of the crowd.

Functional Maths Activity

Air traffic

The following table shows data about air traffic between the UK and abroad from 1980 to 1990.

	1980	1982	1984	1986	1988	1990
Flights (thousands)	507	511	566	616	735	819
Number of passengers (millions)	43	44	51	52	71	77

1 Draw a suitable diagram to illustrate the changes over this ten-year period.

2 Research similar figures from the Internet for the year 2000 and draw a suitable diagram to show the changes that have occurred over the years 1980, 1990 and 2000.

3 Estimate the number of flights and number of passengers for the year 2010.

Algebra: Real-life graphs

10.1 Straight-line distance–time graphs

HOMEWORK 10A

FM 1 Joe was travelling in his car to meet his girlfriend. He set off from home at 9.00 am, and stopped on the way for a break. This distance–time graph illustrates his journey.

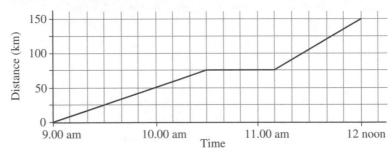

a At what time did he:
   i stop for his break   **ii** set off after his break   **iii** get to his meeting place?
b At what average speed was he travelling:
   i over the first hour   **ii** over the last hour   **iii** for his whole journey?

FM 2 A taxi set off from Hellaby to pick up Jean. It then went on to pick up Jean's parents. It then travelled further, dropping them all off at a shopping centre. The taxi travelled a further 10 km to pick up another party and took them back to Hellaby. This distance–time graph illustrates the journey.

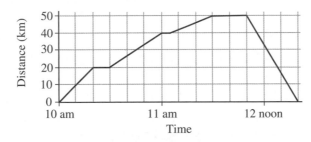

a How far from Hellaby did Jean's parents live?
b How far from Hellaby is the shopping centre?
c What was the average speed of the taxi while only Jean was in the taxi?
d What was the average speed of the taxi back to Hellaby?

FM 3 Grandad took his grandchildren out for a trip. He set off at 1 pm and travelled, for half an hour, away from Sheffield at the average speed of 60 km/h. They stopped to look at the sea and have an ice cream. At two o'clock, they set off again, travelling for a quarter of an hour at a speed of 80 km/h. Here they stopped to play on the sand for half an hour. Grandad then drove the grandchildren back home, at an average speed of 50 km/h. Draw a travel graph to illustrate this story. Use a horizontal axis to represent the time from 1 pm to 5 pm, and a vertical scale from 0 km to 50 km.

FM Functional Maths   **AU** (AO2) Assessing Understanding   **PS** (AO3) Problem Solving

AU **4** A runner sets off at 8 am from point P to jog along a trail at a steady pace of 12 km/h.

One hour later, a cyclist sets off from P on the same trial at a steady pace of 24 km/h. After 30 minutes, the cyclist gets a puncture, which takes 30 minutes to fix. She then sets of at a steady pace of 24 km/h. At what time did the cyclist catch up with the runner? You may use a grid to help you solve this question.

HINTS AND TIPS

This question can be done by many methods, but doing a distance-time graph is the easiest. Mark a grid with a horizontal axis from 8 am to 12 noon and the vertical axis as distance from 0 to 40. Draw lines for both runner and cyclist. Remember that the cyclist doesn't start until 9 am.

HOMEWORK 10B

1 Calculate the gradient of each line.

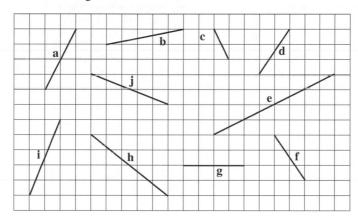

2 Calculate the average speed of the journey represented by each line in the following graphs. The gradient of each line is the speed.

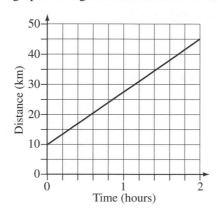

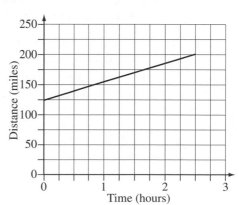

D

3 This is a conversion graph between ounces and grams.
   a Calculate the gradient of the line.
   b Use the graph to find the number of grams equivalent to 1 ounce.

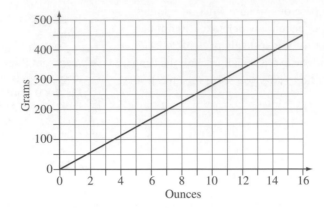

C

FM 4 This graph shows the journey of a car from London to Brighton and back again.

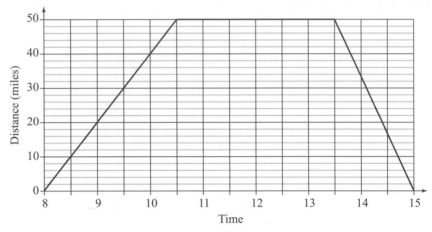

The car leaves at 8 am and returns at 3 pm.
   a For how long does the car stop in Brighton?
   b Was the car travelling faster on the way to Brighton or on the way back to London?
      Explain how you can tell this from the graph.

PS 5 The diagram shows the safe position for a ladder.
Use a grid like the one below and a protractor to work out the
approximate safe angle between the ladder and the ground,
marked x on the diagram.

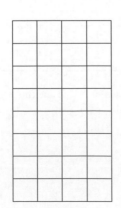

10.2 Other types of graphs

FM 1   The diagram shows the velocity of
a car over 10 seconds.
Calculate the acceleration
a   over the first 2 seconds
b   after 6 seconds.

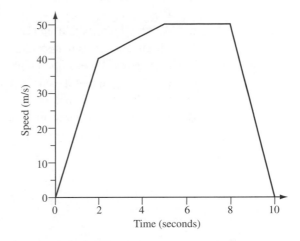

FM 2   The diagram shows the velocity–time graph for a short train journey between stops.
Find:
a   the acceleration over the first 10 seconds
b   the deceleration over the last 20 seconds.

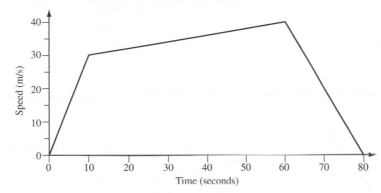

PS 3   Starting from rest (zero velocity), a car travels as indicated below.
- Accelerates at a constant rate over 10 seconds to reach 20 m/s.
- Keeps this velocity for 30 seconds.
- Accelerates over the next 10 seconds to reach 30 m/s.
- Steadily slows down to reach rest (zero velocity) over the next 20 seconds.

a   Draw the velocity–time graph.
b   Calculate the deceleration over the last 20 seconds.

PS 4 This graph represents the journey of a train.

a What was the average speed of the train from A to B?

b How long did the train wait at B?

c Another train starts from C at 11 am and travels non-stop to A at an average speed of 60 mph. Draw the graph of its journey on the copy of the graph.

d Write down how far from A the trains were when they passed each other.

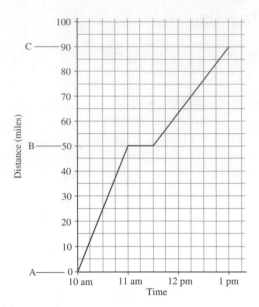

Functional Maths Activity

Driving in the United States

a Petrol is sold in gallons in the United States of America and in litres in the United Kingdom. Until recently, petrol was sold in Imperial gallons in the United Kingdom.

These two graphs show the conversions between US gallons and Imperial gallons, and between Imperial gallons and litres.

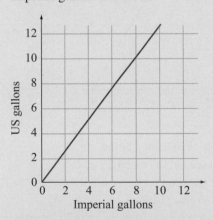

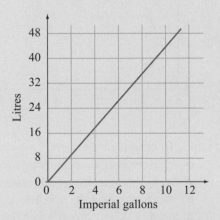

Approximately how many litres are there in a US gallon?

Functional Maths Activity (continued)

b   This is an American advert for a car.

This is a map of the highway distances between four cities in the United States.

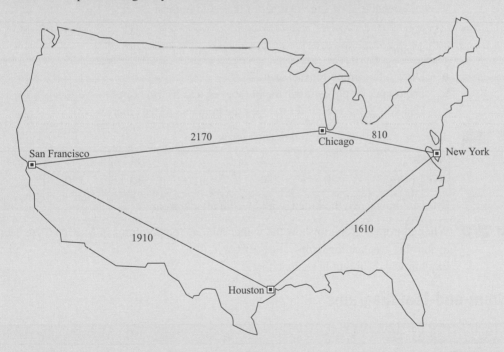

In 2009, Jeff went on a road trip from New York to Chicago, to San Francisco, to Houston and back to New York.

At that time, a US gallon of petrol cost $2.12 on average, while a litre of petrol in the UK cost £0.96p on average.

The exchange rate between pounds and dollars was £1 = $1.60.

Work out:

i   How many litres of fuel were used on this journey.

ii How much more it would have cost at UK prices than at US prices for the fuel.

11 Statistics: Statistical representation

11.1 Line graphs

1 The table shows the estimated number of visits to the cinema in Sheffield.

Year	1970	1975	1980	1985	1990	1995	2000	2005
No. of visits (thousands)	280	110	180	330	510	620	750	810

a Draw a line graph for this data.

b From your graph estimate the number of visits to the cinema in Sheffield in 1998.

c In which 5 year period did the number of visits increase the most?

d Explain the trend in the number of visits. What reasons can you give to explain this trend?

FM 2 Maria started a tea shop and was interested in how trade was picking up over the first few weeks. The table shows the number of teas sold in these weeks.

Week	1	2	3	4	5
Teas sold	67	82	100	114	124

a Draw a line graph for this data.

b From your graph, estimate the number of teas Maria hopes to see in week 6.

c Can you give a reason for the way the numbers of teas increases?

PS 3 A kitten is weighed at the end of each week as:

Week	1	2	3	4	5
Weight (g)	420	480	530	560	580

Estimate how much the kitten would weigh after 8 weeks.

AU 4 When plotting a graph to show the winter midday temperatures in Mexico, Pete decided to start his graph at the temperature 10°C.

Explain why he might have done that.

11.2 Stem-and-leaf diagrams

1 The weights of 15 tomatoes are measured:

62 g, 58 g, 60 g, 48 g, 55 g
53 g, 62 g, 67 g, 57 g, 54 g
60 g, 57 g, 62 g, 47 g, 67 g

a Show the results in an ordered stem-and-leaf diagram, using this key:

6|3 represents 63 g

b What was the greatest weight recorded?

c What was the most common weight measured?

d What is the difference between the largest and smallest weights measured?

2 A group of friends compared how many new DVDs they had watched during the Christmas break.

11, 19, 20, 4, 18, 22, 4, 8, 21, 14, 18, 23, 8, 8, 17, 23

a Show the results in an ordered stem-and-leaf diagram, using this key: 1|3 represents 13 new DVDs watched

b What was the highest number of new DVDs that any of the group watched?

c What was the most common number of new DVDs watched in the break?

FM 3 Chris wanted to know how many people attended a church each Sunday for a month. He recorded the data as:

74, 80, 81, 70, 79, 85, 68, 69, 80, 75, 79, 84, 70, 71, 76, 92, 89, 87, 73, 85

a Show these results in an ordered stem-and-leaf diagram.

b What is the median number of people attending the church?

c What is the range of the numbers?

PS 4 The stem-and-leaf diagram below shows some weights of boys (left) and girls (right) in the same form.

```
          5 | 4 | 2 4 4 6 6 8
  1 4 4 5 7 8 8 | 5 | 0 0 4 5 5 6 7 8
      2 3 4 4 6 | 6 | 3
          1 2 2 | 7 |
```

Explain what the diagram is telling you about the students in the form.

AU 5 The numbers of sweets in a set of packets were each counted, with the following results.

25, 26, 26, 27, 26, 26, 25, 26, 27, 28, 27, 26, 26, 24, 24, 27

Explain why a stem-and-leaf diagram is not a good way to represent this information.

11.3 Scatter diagrams

HOMEWORK 11C

1 The table shows the heights and weights of twelve students in a class.

a Plot the data on a scatter diagram.

b Draw the line of best fit.

c Jayne was absent from the class, but she knows she is 132 cm tall. Use the line of best fit to estimate her weight.

d A new girl joined the class who weighed 55 kg. What height would you expect her to be?

Student	Weight (kg)	Height (cm)
Ann	51	123
Bridie	58	125
Ciri	57.5	127
Di	62	128
Emma	59.5	129
Flo	65	129
Gill	65	133
Hanna	65.5	135
Ivy	71	137
Joy	75.5	140
Keri	70	143
Laura	78	145

2 The table shows the marks for ten pupils in their mathematics and music examinations.

a Plot the data on a scatter diagram. Take the *x*-axis for the mathematics scores and mark it from 20 to 100. Take the *y*-axis for the music scores and mark it from 20 to 100.

b Draw the line of best fit.

c One of the pupils was ill when they took the music examination. Which pupil was it most likely to have been?

d Another pupil, Kris, was absent for the music examination but scored 45 in mathematics, what mark would you expect him to have got in music?

Pupil	Maths	Music
Alex	52	50
Ben	42	52
Chris	65	60
Don	60	59
Ellie	77	61
Fan	83	74
Gary	78	64
Hazel	87	68
Irene	29	26
Jez	53	45

e Another pupil, Lex, was absent for the mathematics examination but scored 78 in music, what mark would you expect him to have got in mathematics?

FM 3 The table shows the time taken and distance travelled by a delivery van for 10 deliveries in one day.

Dist (km)	8	41	26	33	24	36	20	29	44	27
Time (min)	21	119	77	91	63	105	56	77	112	70

a Draw a scatter diagram with time on the horizontal axis.

b Draw a line of best fit on your diagram.

c A delivery takes 45 minutes. How many kilometres would you expect the journey to have been?

d How long would you expect a journey of 30 kilometres to take?

PS 4 Harry records the time taken, in hours, and the distance travelled, in miles, for several different journeys.

Time (h)	1	1.6	2.2	2.6	3.2	3.5	4	4.8	5.2
Distance (m)	42	62	86	104	130	105	165	190	210

Estimate the distance travelled for a journey lasting 200 minutes.

AU 5 Describe what you would expect the scatter graph to look like if someone said that it showed positive correlation.

Functional Maths Activity

Buying wine in the UK

This table shows the number of cases of different types of wine bought in the UK since 2005.

	Number of cases (millions)				
Wines	2005	2006	2007	2008	2009
Chardonnay	6	8	8	8	9
Pinot Grigio	2	3	4	5	5
Sauvignon Blanc	3	4	5	5	6
Merlot	4	4	5	4	5
Shiraz	3	3	4	5	5
Cabernet Sauvignon	4	5	4	5	5
Rioja	2	2	2	2	2

Use appropriate statistical diagrams and measures to summarise the data given in the table.
Then write a report about the sales of wines in the UK over these five years.

12 Probability: Probabilities of events

12.1 Experimental probability

1 Kylie takes a ball at random from a bag that contains six red and four white balls. The table shows her results.

Number of draws	10	20	50	100	500
Number of white balls	2	6	18	42	192

    a Calculate the relative frequency of a white ball at each stage that Kylie recorded her results.

    b What is the theoretical probability of taking a white ball from the bag?

    c If Kylie took a ball out of the bag a total of 5000 times, how many white balls would you expect her to get?

2 Jason made a six-sided spinner. To test it he threw it 600 times. The table shows the results.

Number	1	2	3	4	5	6
Total	98	152	85	102	62	101

    a Work out the relative frequency of each number.

    b How many times would you expect each number to occur if the spinner is fair?

    c Do you think that the spinner is fair? Give a reason for your answer.

3 A sampling bottle contains red, white and blue balls. There are 50 balls in the bottle. Evie performs an experiment to see how many balls of each colour are in the bottle. She shakes the sampling bottle and looks at the one ball that can be seen in the plastic top. She keeps a tally of each colour as she sees it and summarises her totals in the following chart after 100, 250, 400 and 500 trials.

    a Calculate the relative frequencies of the balls at each stage to 3 sf.

    b How many of each colour do you think are in the bag? Explain your answer.

Red	White	Blue	Total
31	52	17	100
68	120	62	250
102	203	95	400
127	252	121	500

4 Which of these methods would you use to estimate or state the probability of each of the events **a** to **g**?

    Method A: Equally likely outcomes   Method B: Survey or experiment

    Method C: Look at historical data

    a There will be an earthquake in Japan.

    b The next person to walk through the door will be female.

    c A Premier League team will win the FA Cup.

    d You will win a raffle.

    e The next car to drive down the road will be foreign.

    f You will have a Maths lesson this week.

    g A person picked at random from your school will go abroad for their holiday.

5 A bag contains a number of counters. Each counter is coloured red, blue or yellow. Each counter is numbered 1 or 2. The table shows the probability of the colour and number for those counters.

Number of counter	Colour of counter		
	Red	Blue	Yellow
1	0.2	0.3	0.1
2	0.2	0.1	0.1

a A counter is taken from the bag at random. What is the probability that:
   i it is red **and** numbered 2   **ii** it is blue **or** numbered 2
   iii it is red **or** numbered 2?

b There are two yellow counters in the bag. How many counters are in the bag altogether?

6 A check was done one week to see how many passengers on the Number 79 bus were pensioners:

	Mon	Tue	Wed	Thu	Fri
Passengers	950	730	1255	796	980
Pensioners	138	121	168	112	143

For each day, what is the probability that a pensioner was the 400th passenger to board the bus that day?

PS 7 Andrew made a six-sided spinner.

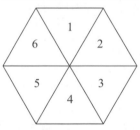

He tested it to see if it was fair.
He spun the spinner 240 times and recorded the results in a table.

Number spinner lands on	1	2	3	4	5	6
Frequency	43	38	31	41	42	44

Do you think the spinner is fair?
Give reasons for your answer.

AU 8 Aleena tossed a coin 50 times to see how many tails she would get.
She said, 'If this is a fair coin, then I should get 25 tails.'
Explain why she is wrong.

12.2 Mutually exclusive and exhaustive events

HOMEWORK 12B

1 Say whether these pairs of events are mutually exclusive or not.
  a Tossing two heads with two coins/tossing two tails with two coins.
  b Throwing an even number with a dice/throwing an odd number with a dice.
  c Drawing a Queen from a pack of cards/drawing an Ace from a pack of cards.
  d Drawing a Queen from a pack of cards/drawing a red card from a pack of cards.
  e Drawing a red card from a pack of cards/drawing a Heart from a pack of cards.

2 Which of the pairs of mutually exclusive events in Question **1** are also exhaustive?

3 A letter is to be chosen at random from this set of letter-cards.

| M | I | S | S | I | S | S | I | P | P | I |

  a What is the probability that the letter chosen is:
     i an S   **ii** a P   **iii** a vowel?
  b Which of these pairs of events are mutually exclusive?
     i Picking an S / picking a P.   **ii** Picking an S / picking an I.
     iii Picking an I / picking a consonant.
  c Which pair of mutually exclusive events in part **b** is also exhaustive?

FM 4 Two people are to be chosen for a job from these six people.
  Ann   Joan   Jack   John   Arthur   Ethel
  a List all of the possible pairs (there are 15 altogether).
  b What is the probability that the pair of people chosen will:
     i both be female   **ii** both be male   **iii** both have the same initial
     iv have different initials?
  c Which of these pairs of events are mutually exclusive?
     i Picking two women / picking two men.
     ii Picking two people of the same sex / picking two people of opposite sex.
     iii Picking two people with the same initial / picking two men.
     iv Picking two people with the same initial / picking two women.
  d Which pair of mutually exclusive events in part **c** is also exhaustive?

5 For breakfast I like to have toast, porridge or cereal. The probability that I have toast is $\frac{1}{3}$, the probability that I have porridge is $\frac{1}{2}$. What is the probability that I have cereal?

6 A person is chosen at random. Here is a list of events.
     Event A: the person chosen is male   Event B: the person chosen is female
     Event C: the person chosen is over 18   Event D: the person chosen is under 16
     Event E: the person chosen has a degree   Event F: the person chosen is a teacher
  For each of the pairs of events **i** to **x**, say whether they are:
  a mutually exclusive   **b** exhaustive
  If they are not mutually exclusive, explain why.
     i Event A and Event B   **ii** Event A and Event C
     iii Event B and Event D   **iv** Event C and Event D
     v Event D and Event F   **vi** Event E and Event F
     vii Event E and Event D   **viii** Event A and Event E
     ix Event C and Event F   **x** Event C and Event E

7 An amateur weather man, Steve, records the weather over a year in his village. He knows that the probability of a windy day is 0.4 and that the probability of a rainy day is 0.6. Steve says 'This means it will be either rainy or windy each day as 0.4 + 0.6 = 1, which is certain.' Explain why Steve is wrong.

PS 8 Four brothers, David, Malcolm, Brian and Kevin, regularly run races against each other in the park.
The chances of:
David winning the race is 0.3.
Malcolm winning the race is $\frac{1}{5}$.
Brian winning the race is 45%.
What is the chance of Kevin winning the race?

AU 9 Gareth will either walk, go by bus or be given a lift by his dad to get to football training.
If he walks, the probability that he is late for the practice is 0.4.
If he goes by bus, the probability that he is late for the practice is 0.5.
Explain why it is not necessarily true that, if his dad gives him a lift, his probability of being late for school is 0.1.

12.3 Expectation

HOMEWORK 12C

1 I throw an ordinary dice 600 times. How many times can I expect to get a score of 1?

2 I toss a coin 500 times. How many times can I expect to get a tail?

3 I draw a card from a pack of cards and replace it. I do this 104 times. How many times would I expect to get:
a a red card **b** a Queen **c** a red seven
d the Jack of Diamonds?

4 The ball in a roulette wheel can land on any number from 0 to 36. I always bet on the same block of numbers 1–6. If I play all evening and there is a total of 111 spins of the wheel in that time, how many times could I expect to win?

5 I have five tickets for a raffle and I know that the probability of my winning the prize is 0.003. How many tickets were sold altogether in the raffle?

6 In a bag there are 20 balls, ten of which are red, three yellow, and seven blue. A ball is taken out at random and then replaced. This is repeated 200 times. How many times would I expect to get:
a a red ball **b** a yellow or blue ball
c a ball that is not blue **d** a green ball?

7 A sampling bottle contains black and white balls. It is known that the probability of getting a black ball is 0.4. How many white balls would you expect to get in 200 samples if one ball is sampled each time?

8 **a** Fred is about to take his driving test. The chance he passes is $\frac{1}{3}$.
His sister says 'Don't worry if you fail because you are sure to pass within three attempts because $3 \times \frac{1}{3} = 1$'. Explain why his sister is wrong.
b If Fred does fail would you expect the chance that he passes next time to increase or decrease? Explain your answer.

FM 9 Kara rolls two dice 200 times.
a How many times would she expect to roll a double?
b How many times would she expect to roll a total score greater than 7?

10 An opinion poll used a sample of 200 voters in one area. 112 said they would vote for Party A. There are a total of 50 000 voters in the area.

a If they all voted, how many would you expect to vote for Party A?

b The poll is accurate within 10%. Can Party A be confident of winning?

PS 11 A roulette wheel has 37 spaces for the ball to land on. The spaces are numbered 0 to 36. I always bet on a prime number.

If I play the game all evening and anticipate playing 100 times, how many times would I expect to win on the roulette table?

AU 12 A headteacher is told that the probability of any student being left-handed is 0.14. How will she find out how many of her students she should expect to be left-handed?

12.4 Two-way tables

HOMEWORK 12D

1 The two-way table shows the number of children and the number of computers in 40 homes in one road in a town.

	Number of children			
Number of computers	0	1	2	3
0	3	0	0	0
1	4	10	2	2
2	0	4	9	3
3	0	0	2	1

a How many homes have exactly two children and two computers?

b How many homes altogether have two computers?

c What percentage of the homes have two computers?

d What percentage of the homes with just one child have one computer?

FM 2 The two-way table shows the part-time earnings of a set of students during one summer break.

Earnings per week	Male	Female
£0 < E ≤ £50	4	1
£50 < E ≤ £100	4	1
£100 < E ≤ £150	11	4
£150 < E ≤ £200	24	14
£200 < E ≤ £250	16	10
More than £250	2	1

a What percentage of the male students earned between £100 and £150 per week?

b What percentage of the female students earned between £100 and £150 per week?

c Estimate the mean earnings of this set of students.

d Which sex has the greater estimated mean earnings? Explain how you might do this without doing the actual calculation.

3 Here are two fair spinners.

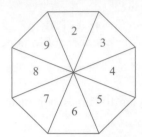 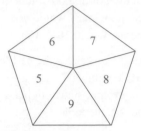

The spinners are spun and the two numbers are added together.

a Draw a probability sample space diagram.

b What is the most unlikely score?

c What is the probability of getting a total of 10?

d What is the probability of getting a total of 9 or more?

e What is the probability of getting a total that is an even number?

PS 4 Two six-sided spinners are spun.

Spinner X has the numbers 2, 4, 6, 8, 9 and 11.

Spinner Y has the numbers 3, 4, 5, 6, 7, and 8.

What is the probability that when the two spinners are spun, the two numbers given will multiply to a total greater than 30?

AU 5 Connie planted some tomato plants and kept them in the kitchen, while her husband Harold planted some in the garden.

After the summer, they compared their tomatoes.

	Connie	Harold
Mean diameter	1.9 cm	4.3 cm
Mean number of tomatoes per plant	23.2	12.3

Use the data in the table to explain who had the better crop of tomatoes.

12.5 Addition rule for events

HOMEWORK 12E

1 Shaheeb throws an ordinary dice. What is the probability that he throws:

a an even number **b** 5 **c** an even number or 5?

2 Jane draws a card from a pack of cards. What is the probability that she draws:

a a red card **b** a black card **c** a red or a black card?

3 Natalie draws a card from a pack of cards. What is the probability that she draws one of the following:

a Ace **b** King **c** Ace or King?

4 A letter is chosen at random from the letters in the word STATISTICS. What is the probability that the letter will be:

a S **b** a vowel **c** S or a vowel?

5 A bag contains 10 white balls, 12 black balls and eight red balls. A ball is drawn at random from the bag. What is the probability that it will be:

a white **b** black **c** black or white

d not red **e** not red or black?

6 A spinner has numbers and colours on it, as shown in
the diagram. Their probabilities are given in the tables.

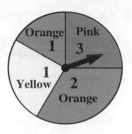

Orange	0.5
Yellow	0.25
Pink	0.25

1	0.4
2	0.35
3	0.25

When the spinner is spun what is the probability of each of
the following?

a Orange or pink **b** 2 or 3   **c** 3 or pink   **d** 2 or yellow

e **i** Explain why the answer to **c** is 0.25 and not 0.5.

  ii What is the answer to P(2 or orange)?

7 Debbie has 10 CDs in her multi-changer, four of which are rock, two are dance and four
are classical. She puts the player on random play. What is the probability that the first CD
will be:

a rock or dance   **b** rock or classical **c** not rock?

8 Frank buys one dozen free-range eggs. The farmer tells him that a quarter of the eggs his
hens lay have double yolks.

a How many eggs with double yolks can Frank expect to get?

b He cooks three and finds they all have a single yolk. He argues that he now has a
1 in 3 chance of a double yolk from the remaining eggs. Explain why he is wrong.

9 John has a bag containing six red, five blue and four green balls. One ball is picked from
the bag at random. What is the probability that the ball is:

a red or blue   **b** not blue   **c** pink   **d** red or not blue?

10 Neil and Mandy put music onto their mp3 player so that at their party they have a variety
of background music. They put 100 different tracks onto the player as follows: 10 love
songs, 15 rap tracks, 35 rock tracks and 40 contemporary tracks.
They set it to play the tracks at random and continuously.

a What is the probability that:

  i the first track played is a love song?

  ii the last track of the evening is either rock or a contemporary track?

  iii the track when they start eating is not a rap track?

b At the party, at midnight, they want to announce their engagement. They want to
have a love song or a contemporary track playing. What is the probability that they
will not get a track of their choice?

c The party lasts for six hours. For how much time, in hours and minutes, would you
expect the mp3 player to have been playing rock tracks?

PS 11 Joy, Vicky and Max play cards together every Sunday night. Joy is always the favourite to
win with a probability of 0.65.
In the year 2009 there were 52 Sundays and Vicky won 10 times.
How many times in the year would you expect Max to have won?

AU 12 Kathy has blue, black, brown and yellow jumpers in her bedroom wardrobe.
Brian is asked to bring a jumper down to her.
Why is P(neither blue nor yellow) not equal to P(not blue) + P(not yellow)?

12.6 Combined events

HOMEWORK 12F

1 Two dice are thrown together. Draw a probability diagram to show the total score.
  a What is the probability of a score that is:
    i 7 **ii** 5 or 8 **iii** bigger than 9 **iv** between 2 and 5
    v odd **vi** a non-square number?

2 Two dice are thrown. Draw a probability diagram to show the outcomes as pairs of co-ordinates.
  What is the probability that:
  a the score is a 'double'
  b at least one of the dice shows 3
  c the score on one dice is three times the score on the other dice
  d at least one of the dice shows an odd number
  e both dice show a 5
  f at least one of the dice will show a 5
  g exactly one dice shows a 5?

3 Two dice are thrown. The score on one dice is doubled and the score on the other dice is subtracted.
  Complete the probability space diagram.
  For the event described above, what is the probability of a score of:
  a 1
  b a negative number
  c an even number
  d 0 or 1
  e a prime number

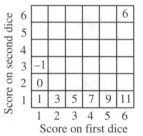

4 When two coins are tossed together, what is the probability of:
  a two heads or two tails **b** a head and a tail **c** at least one head?

5 When three coins are tossed together, what is the probability of:
  a three heads or three tails **b** two tails and one head **c** at least one head?

6 When a dice and a coin are thrown together, what is the probability of each of the following outcomes?
  a You get a tail on the coin and a 3 on the dice.
  b You get a head on the coin and an odd number on the dice.

7 Max buys two bags of bulbs from his local garden centre. Each bag has four bulbs. Two bulbs are daffodils, one is a tulip and one is a hyacinth. Max takes one bulb from each bag.

Hyacinth				HH
Tulip	DT			
Daffodil				
Daffodil	DD	DD	TD	
	Daffodil	Daffodil	Tulip	Hyacinth

  a There are six possible different pairs of bulbs. List them all.
  b Complete the sample space diagram.
  c What is the probability of getting two daffodil bulbs?
  d Explain why the answer is not $\frac{1}{6}$.

FM 8 Shehab walked into his local supermarket and saw a competition:

> Roll 2 dice.
>
> Get a total of 2
>
> and win a £20 note.
>
> Only 50p a go.

Shehab thought about having a go.

a Draw the sample space for this event.

b What is the probability of winning a £20 note?

c How many goes should he have in order to expect to have won at least once?

d If he had 100 goes, how many times could he expect to have won?

PS 9 I toss four coins. What is the probability that I will get more heads than tails?

AU 10 I roll a dice four times and add the four numbers showing.

Explain the difficulty in drawing a sample space to show all the possible events.

Functional Maths Activity

Lottery competition

A company runs its own version of a lottery but only using the numbers 1 to 20. Four numbers are chosen each week.

1 Andrew chooses the numbers, 1, 2, 3 and 4.
What is his probability of winning the lottery?

2 Evie says Andrew is unlikely to win with four consecutive numbers. She chooses the ages of her nieces and nephews which are 5, 11, 12 and 16.
She says, 'I have a better chance than Andrew.'
Is she correct? Explain your answer.

3 There are 870 employees in the company. Every week they all choose their four numbers and pay 10 pence.
If their four numbers are selected they win £50. There are no rollovers.
In one year, the lottery is run 32 times.
Any profit made goes to a local charity. How much money would the company expect to raise for charity in a year?

Algebra: Number and sequences

13.1 Number sequences

HOMEWORK 13A

1 Look at the following number sequences. Write down the next three terms in each and try to explain the pattern.

 a   4, 6, 8, 10, …   **b**   3, 6, 9, 12, …   **c**   2, 4, 8, 16, …
 d   5, 12, 19, 26, …   **e**   3, 30, 300, 3000, …   **f**   1, 4, 9, 16, …

2 Look carefully at each number sequence below. Find the next two numbers in the sequence and try to explain the pattern.

 a   1, 1, 2, 3, 5, 8, 13, 21, …   **b**   2, 3, 5, 8, 12, 17, …

3 Look at the sequences below. Find the rule for each sequence and write down its next three terms.

 a   7, 14, 28, 56, …   **b**   3, 10, 17, 24, 31, …   **c**   1, 3, 7, 15, 31, …
 d   40, 39, 37, 34, …   **e**   3, 6, 11, 18, 27, …   **f**   4, 5, 7, 10, 14, 19, …
 g   4, 6, 7, 9, 10, 12, …   **h**   5, 8, 11, 14, 17, …   **i**   5, 7, 10, 14, 19, 25, …
 j   10, 9, 7, 4, …   **k**   200, 40, 8, 1.6, …   **l**   3, 1.5, 0.75, 0.375, …

FM 4 A physiotherapist uses the formula below for charging for a series of n sessions when they are paid for in advance.

For $n \leq 5$, cost will be £$(35n + 20)$

For $6 \leq n \leq 10$, cost will be £$(35n + 10)$

For $n \geq 11$, cost will be £$35n$

 a   How much will the physiotherapist charge for eight sessions booked in advance?
 b   How much will the physiotherapist charge for 14 sessions booked in advance?
 c   One client paid £220 in advance for a series of sessions.
    How many sessions did she book?
 d   A runner has a leg injury and is not sure how many sessions it will take to cure.
    The runner books four sessions in advance, and after the sessions starts to run in
    races again. The leg injury returns and he has to book another three sessions before
    he is finally cured.
    How much more did it cost him than if he had booked seven sessions in advance?

PS 5 The formula for working out a series of fractions is

$$\frac{n + 2}{2n + 1}$$

Show that in the first eight terms only one of the fractions is a terminating decimal.

HINTS AND TIPS

If you set this up on a spreadsheet, find the relationship between the denominators of the terms that give terminating decimals in this series.

13.2 Finding the *n*th term of a linear sequence

HOMEWORK 13B

1 Find the *n*th term in each of these linear sequences.
  a 5, 7, 9, 11, 13 …   **b** 3, 11, 19, 27, 35, …   **c** 6, 11, 16, 21, 26, …
  d 3, 9, 15, 21, 27, …   **e** 4, 7, 10, 13, 16, …   **f** 3, 10, 17, 24, 31, …

2 Find the 50th term in each of these linear sequences.
  a 3, 5, 7, 9, 11, …   **b** 5, 9, 13, 17, 21, …   **c** 8, 13, 18, 23, 28, …
  d 2, 8, 14, 20, 26, …   **e** 5, 8, 11, 14, 17, …   **f** 2, 9, 16, 23, 30, …

3 For each sequence **a** to **f**, find
  i the *n*th term   **ii** the 100th term   **iii** the term closest to 100
  a 4, 7, 10, 13, 16, …   **b** 7, 9, 11, 13, 15, …   **c** 3, 8, 13, 18, 23, …
  d 1, 5, 9, 13, 17, …   **e** 2, 10, 18, 26, …   **f** 5, 6, 7, 8, 9, …

4 A sequence of fractions is $\frac{3}{5}, \frac{5}{8}, \frac{7}{11}, \frac{9}{14}, \dots$
  a Find the *n*th term in the sequence.   **b** Change each fraction to a decimal.
  c What, as a decimal, will be the value of the:
    i 100th term   **ii** 1000th term?
  d Use your answers to part **c** to predict what the 10 000th term and the millionth term are. (Check these out on your calculator.)

5 **a** A number pattern begins 1, 1, 2, 3, 5, 8, …
    i What is the next number in this pattern?
    ii The number pattern is continued. Explain how you would find the 10th number in the pattern.
  b Another number pattern begins 1, 5, 9, 13, 17, … . Write down, in terms of *n*, the *n*th term in this pattern.

FM 6 This chart is used by a taxi firm for the charges for journeys of *k* kilometres.

k	1	2	3	4	5	6	7	8	9	10
Charge (£)	4.50	6.50	8.50	10.50	12.50	15.00	17.00	19.00	21.00	23.00
k	11	12	13	14	15	16	17	18	19	20
Charge (£)	26.00	28.00	30.00	32.00	34.00	37.00	39.00	41.00	43.00	45.00

  a Using the charges for 1 to 5 km, work out an expression for the *k*th term.
  b Using the charges for 6 to 10 km, work out an expression for the *k*th term.
  c Using the charges for 10 to 15 km, work out an expression for the *k*th term.
  d Using the charges for 16 to 20 km, work out an expression for the *k*th term.
  e What is the basic charge per kilometre?

PS 7 A series of fractions is $\frac{3}{7}, \frac{5}{10}, \frac{7}{13}, \frac{9}{16}, \frac{11}{19}, \dots$
  a Write down an expression for the *n*th term of the numerators.
  b Write down an expression for the *n*th term of the denominators.
  c **i** Work out the fraction when *n* = 1000.
    ii Give the answer as a decimal.
  d Will the terms of the series ever be greater than $\frac{2}{3}$?
    Explain your answer.

13.3 Special sequences

HOMEWORK 13C

PS 1 p is an even number, q is a square number.
State if the following are odd or even or could be either odd or even.

a $p + 1$ **b** $p + q$ **c** $2q$ **d** $3p - 1$
e p^2 **f** $pq - 1$ **g** $p^2 + 4q$ **h** $(p + q)(p - q)$

2 The powers of 3 are $3^1, 3^2, 3^3, 3^4, \dots$
This gives the sequence 3, 9, 27, 81, ...

  a Continue the sequence for another three terms.

  b The nth term is given by 3^n.
     Give the nth terms of each of these sequences.

     i 2, 8, 26, 80, ...

     ii 6, 18, 54, 162, ...

3 The negative powers of 10, starting with the power 0, are:
$10^0, 10^{-1}, 10^{-2}, 10^{-3}, \dots$
This gives the sequence 0, 0.1, 0.01, 0.001

  a Describe the connection between the numerical value of the power and the number of zeros after the decimal point.

  b If $10^{-n} = 0.0000001$, what is the value of n?

4 **a** Complete the table below to show the possible outcomes of adding a prime number to odd and even numbers, showing whether the answer is always odd, always even or could be either odd or even.

+	Prime	Odd	Even
Prime	Either		
Odd		Even	
Even			Even

  b Complete the table below to show the possible outcomes of multiplying a prime number to odd and even numbers, showing whether the answer is always odd, always even or could be either odd or even.

+	Prime	Odd	Even
Prime	Either		
Odd		Odd	
Even			Even

PS 5 **a** Draw an equilateral triangle with each side 9 cm.
The perimeter will be 27 cm.

  b Draw another equilateral triangle of side 3 cm on each edge.
Work out the perimeter of the new shape.

  c Draw another equilateral triangle of side 1 cm on each of the remaining edges.
Work out the perimeter of the new shape.

  d The next step would be to draw a triangle of side $\frac{1}{3}$ cm on each remaining edge, but this will be difficult to draw.
You should be able to write down the perimeter using the pattern of the perimeters so far.

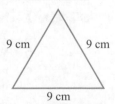

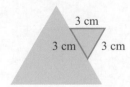

The formula is $27 \times \left(\dfrac{4}{3}\right)^{n-1}$

You will need a calculator with a power button (^).

Work out $27 \times (4 \div 3) \wedge 0$, which should equal 27.

Then work out $27 \times (4 \div 3) \wedge 1$, which should equal your answer to the perimeter in part **b**.

Use the formula to work out the perimeter of the next drawing when $n = 4$

e Work out the perimeter when $n = 100$

If we kept on drawing triangles, the perimeter would become infinite.

This is an example of a shape that has a finite area surrounded by an infinite perimeter.

13.4 General rules from given patterns

HOMEWORK 13D

1 A pattern of shapes is built up from matchsticks as shown.

  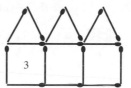

a Draw the fourth diagram.

b How many matchsticks are in the nth diagram?

c How many matchsticks are in the 25th diagram?

d With 200 matchsticks, which is the biggest diagram that could be made?

2 A pattern of hexagons is built up from matchsticks.

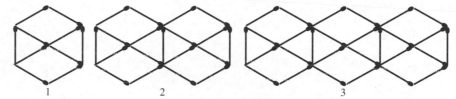

a Draw the fourth set of hexagons in this pattern.

b How many matchsticks are needed for the nth set of hexagons?

c How many matchsticks are needed to make the 60th set of hexagons?

d If there are only 100 matchsticks, which is the largest set of hexagons that could be made?

3 A conference centre had tables each of which could sit three people. When put together, the tables could seat people as shown.

 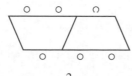 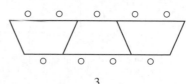

a How many people could be seated at four tables?

b How many people could be seated at n tables put together in this way?

c A conference had 50 people who wished to use the tables in this way. How many tables would they need?

PS **4** The picture shows a pattern of cards.

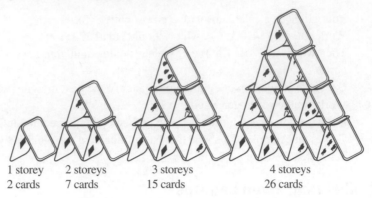

1 storey   2 storeys   3 storeys   4 storeys
2 cards   7 cards   15 cards   26 cards

a The four-storey house of cards is to be made into a five-storey house of cards. How many more cards are needed?

b Look at the sequence 2, 7, 15, 26, …

   i Calculate the sixth term in this sequence.

   ii Explain how you found your answer.

5 **a** Find the first three terms that these two sequences have in common:

       2, 5, 8, 11, 14, ...

       3, 7, 11, 15, 19, ...

b Write down the nth term of the sequence that is the answer to part **a**.

Functional Maths Activity

Packaging

To tie up a cuboidal package that is L cm long, W cm wide and H cm high, this formula gives the length of string needed:

$S = 2L + 2W + 4H + 20$

Masood wants to send eight identical cuboidal boxes with sides of 15 cm in one package.

He can arrange them in three different ways to make a cuboidal shape.

Two of these ways are shown below.

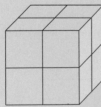 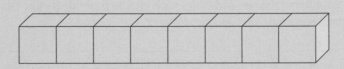

Which of the three ways of arranging the eight cubes in a cuboid will use the least amount of string?

Algebra: Graphs and their equations

14.1 Linear graphs

Draw the graph for each of the equations given.

Follow these hints.

- Use the highest and smallest values of x given as your range.
- When the first part of the function is a division, pick x-values that divide exactly to avoid fractions.
- Always label your graphs. This is particularly important when you are drawing two graphs on the same set of axes.
- Create a table of values. You will often have to complete these in your examinations.

1 Draw the graph of $y = 2x + 3$ for x-values from 0 to 5 ($0 \leqslant x \leqslant 5$)

2 Draw the graph of $y = 3x - 1$ ($0 \leqslant x \leqslant 5$)

3 Draw the graph of $y = \dfrac{x}{2} - 2$ ($0 \leqslant x \leqslant 12$)

4 Draw the graph of $y = 2x + 1$ ($-2 \leqslant x \leqslant 2$)

5 Draw the graph of $y = \dfrac{x}{2} + 5$ ($-6 \leqslant x \leqslant 6$)

6 **a** On the same set of axes, draw the graphs of
$y = 3x - 1$ and $y = 2x + 3$ ($0 \leqslant x \leqslant 5$)
  b Where do the two graphs cross?

7 **a** On the same axes, draw the graphs of
$y = 4x - 3$ and $y = 3x + 2$ ($0 \leqslant x \leqslant 6$)
  b Where do the two graphs cross?

8 **a** On the same axes, draw the graphs of
$y = \dfrac{x}{2} + 1$ and $y = \dfrac{x}{3} + 2$ ($0 \leqslant x \leqslant 12$)
  b Where do the two graphs cross?

9 **a** On the same axes, draw the graphs of
$y = 2x + 3$ and $y = 2x - 1$ ($0 \leqslant x \leqslant 4$)
  b Do the graphs cross? If not, why not?

10 **a** Copy and complete the table to draw the graph of $x + y = 6$ ($0 \leqslant x \leqslant 6$)

x	0	1	2	3	4	5	6
y							

  b Now draw the graph of $x + y = 3$.

D

FM 11 CityCabs uses this formula to work out the cost of a journey of k kilometres:
$C = 2.5 + k$
TownCars uses this formula to work out the cost of a journey of k kilometres:
$C = 2 + 1.25k$

a On the grid below, draw lines to represent these formulae.

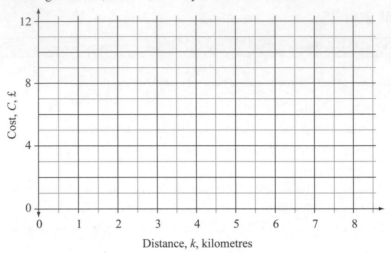

b At what length of journey do CityCabs and TownCars charge the same amount?

C

AU 12 The line $x + y = 5$ is drawn on the grid below.

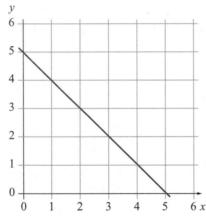

Draw a line of the form $x = a$ **and** a line of the form $y = b$ so that the area between the 3 lines is 4.5 square units.

HOMEWORK 14B

1 Find the gradient of each of these lines.

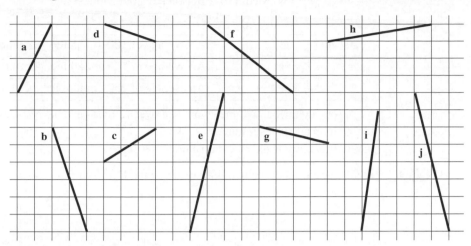

2 Draw lines with gradients of
  a 3   **b** $\frac{1}{2}$ **c** −1   **d** 8   **e** $\frac{3}{4}$ **f** $-\frac{1}{3}$

3 **a** Draw a pair of axes with both x and y showing from −10 to 10.
  b Draw on the grid, lines with the following gradients; start each line at the origin.
    Clearly label each line.
    i $\frac{1}{2}$ **ii** 1 **iii** 2 **iv** 4 **v** −4 **vi** −2 **vii** −1 **viii** $-\frac{1}{2}$
  c Describe the symmetries of your diagram.

FM 4 This graph shows the profile of a fell race. The horizontal axis shows the distance in miles of the race.
The vertical axis is the height above sea level throughout the race.
There are 5280 feet in a mile.

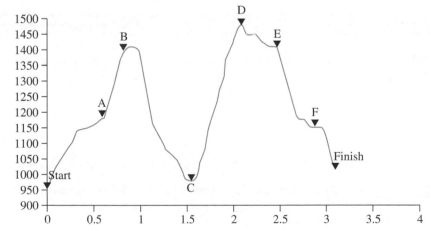

  a Work out the approximate gradient of the race from the start to point A.
  b Work out the approximate gradient from C to D.

c Fell races are classified in terms of distance and amount of ascent.

Distance	S (Short)	Less than 6 miles
	M (Medium)	Between 6 and 12 miles
	L (Long)	Over 12 miles
Ascent	C	An average of 100 to 125 feet per mile
	B	An average of 125 to 250 feet per mile
	A	An average of 250 or more feet per mile

So, for example, an AL race would be over 12 miles and have at least 250 feet of ascent on average per mile.

What category is the race on the previous page?

AU **5** Write the gradients of the two lines below in the form 1 : *n*.

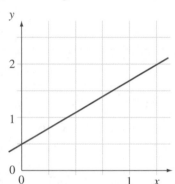 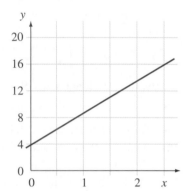

PS **6** Put the following gradients in order of steepness starting with the shallowest.
1 horizontal, 3 vertical   2 horizontal, 9 vertical   3 horizontal, 7 vertical
4 horizontal, 5 vertical   5 horizontal, 7 vertical   6 horizontal, 13 vertical

14.2 Drawing graphs by the gradient-intercept method

HOMEWORK 14C

1 Draw these lines using the gradient-intercept method. Use the same grid, taking both *x* and *y* from −10 to 10. If the grid gets too 'crowded', draw another one.

a $y = 2x + 4$ **b** $y = 3x - 2$ **c** $y = \frac{1}{2}x + 1$ **d** $y = x + 5$

e $y = 4x - 1$ **f** $y = 2x - 5$ **g** $y = \frac{1}{4}x - 1$ **h** $y = x - 1$

i $y = 6x - 2$ **j** $y = x + 3$ **k** $y = \frac{3}{4}x - 2$ **l** $y = 3x - 4$

For the next questions, use grids showing *x* from −6 to 6 and *y* from −8 to 8.

2 **a** Using the gradient-intercept method, draw the following lines on the same grid.
    i $y = 3x + 2$ **ii** $y = 2x - 1$
  b Where do the lines cross?

3 **a** Using the gradient-intercept method, draw the following lines on the same grid.
    i $y = x + 2$ **ii** $y = 5x$
  b Where do the lines cross?

AU 4 Here are 3 lines.

A: $y = 4x - 3$ B: $2y = 8x - 6$ C: $y = 2x - 3$

  a State a mathematical property that lines A and B have in common.

  b State a mathematical property that lines B and C have in common.

  c Which of the following points is the intersection of lines A and C?

    $(1, -3)$ $(0, 3)$ $(0, -3)$ $(1, 3)$

PS 5 Look at these two lines drawn on the same axes:

  a What is the gradient of line A?

  b What is the gradient of line B?

  c What angle is there between line A and B?

  d What relationship do the gradients of A and B have with each other?

  e Another line, C, has a gradient of $-\frac{1}{2}$.

    What is the gradient of a line perpendicular to C?

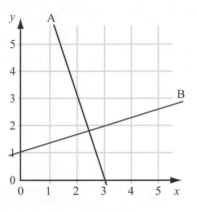

HOMEWORK 14D

1 Draw these lines using the cover-up method. Use the same grid, taking x from -10 to 10 and y from -10 to 10. If the grid gets too 'crowded', draw another.

  a $2x + 3y = 6$ **b** $3x + 4y = 12$ **c** $5x - 4y = 20$ **d** $x + y = 8$

  e $2x - 3y = 18$ **f** $x - y = 6$ **g** $3x - 5y = 15$ **h** $3x - 2y = 12$

  i $5x + 4y = 30$ **j** $x + y = -1$ **k** $x + y = 5$ **l** $x - y = -6$

2 **a** Using the cover-up method, draw the following lines on the same grid.

    i $x + 2y = 4$ **ii** $2x - y = 2$

  b Where do the lines cross?

3 Using the cover-up method, draw the following lines on the same grid.

    $x + 2y = 6$ $2x - y = 2$

  Where do the lines cross?

AU 4 Here are three lines:

A: $3x + 4y = 12$ B: $x - 2y = 3$ C: $x + y = 3$

  a State a mathematical property that lines A and C have in common.

  b State a mathematical property that lines B and C have in common.

  c The line A crosses the x axis at $(4, 0)$.

    The line B crosses the y axis at $(0, -1\frac{1}{2})$

    Find values of a and b so that the line $ax - by = 12$ passes through these two points.

PS 5 The diagram shows a hexagon *ABCDEF*.
The equation of the line through *A* and *B* is $y = 3$.
The equation of the line through *B* and *C* is $x + y = 4$

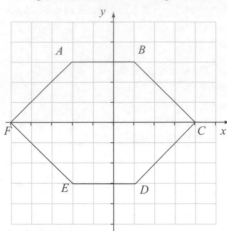

a Write down the equation of the lines through the points:
  i *E* and *D*
  ii *D* and *C*
  iii *A* and *F*
  iv *E* and *F*

b The gradient of the line through *E* and *B* is 2.
Write down the gradient of the line through *A* and *D*.

14.3 Finding the equation of a line from its graph

HOMEWORK 14E

1 Give the equation of each of these lines.

a 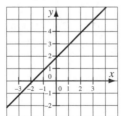 b 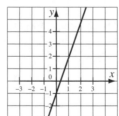 c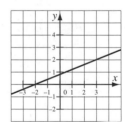

PS 2 In each of these grids, there are two lines.

a 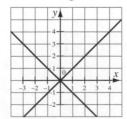 b 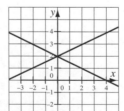 c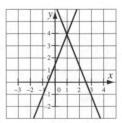

i Find the equation of each line.
ii Describe any symmetries you see about the two lines.

PS 3 The diagram shows four lines crossing to create a rectangle. Write down the equation of each of the four lines.

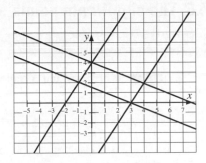

AU 4 A line passes through the points $(-1, 3)$ and $(-2, 5)$.

a Explain how you can tell that the line also passes through $(0, 1)$.

b Explain how you can tell that the line has a gradient of -2.

c Work out the equation of the line that passes through $(-1, 5)$ and $(-2, 8)$.

14.4 Quadratic graphs

HOMEWORK 14F

1 **a** Copy and complete the table or use a calculator to work out values for the graph of $y = 2x^2$ for $-3 \leqslant x \leqslant 3$.

x	-3	-2	-1	0	1	2	3
$y = 2x^2$	18		2			8	

b Use your graph to find the value of y when $x = -1.4$.

c Use your graph to find the values of x that give a y-value of 10.

2 **a** Copy and complete the table or use a calculator to work out values for the graph of $y = x^2 + 3$ for $-5 \leqslant x \leqslant 5$.

x	-5	-4	-3	-2	-1	0	1	2	3	4	5
$y = x^2 + 3$	28		12					7			28

b Use your graph to find the value of y when $x = 2.5$.

c Use your graph to find the values of x that give a y-value of 10.

3 **a** Copy and complete the table or use a calculator to work out values for the graph of $y = x^2 - 3x + 2$ for $-3 \leqslant x \leqslant 4$.

x	-3	-2	-1	0	1	2	3	4
$y = x^2 - 3x + 2$	20			2			2	

b Use your graph to find the value of y when $x = -1.5$.

c Use your graph to find the values of x that give a y-value of 2.5.

PS 4 Tom is drawing quadratic equations of the form $y = x^2 + bx + c$.

He notices that two of his graphs pass through the point $(2, 5)$.

Which of the following equations are those of the two graphs?

Equation A: $y = x^2 + 3$

Equation B: $y = x^2 + 1$

Equation C: $y = x^2 + 2x - 3$

Equation D: $y = x^2 - x + 5$

14.5 The significant points of a quadratic graph

HOMEWORK 14G

1 a Copy and complete the table to draw the graph of $y = x^2 - 5x + 4$ for $-1 \leqslant x \leqslant 6$

x	-1	0	1	2	3	4	5	6
$y = x^2 - 5x + 4$	10	4				0		

  b Use your graph to find the roots of the equation $x^2 - 5x + 4 = 0$

2 a Copy and complete the table to draw the graph of $y = x^2 - 3x + 2$ for $-1 \leqslant x \leqslant 5$

x	-1	0	1	2	3	4	5
$y = x^2 - 3x + 2$	6	2			1		

  b Use your graph to find the roots of the equation $x^2 - 3x + 2 = 0$

3 a Copy and complete the table to draw the graph of $y = x^2 + 4x - 6$ for $-5 \leqslant x \leqslant 2$

x	-5	-4	-3	-2	-1	0	1	2
$y = x^2 + 4x - 6$	-1							6

  b Use your graph to find the roots of the equation $x^2 + 4x - 6 = 0$

4 Using your answers to question **1**, write down:
  a the coordinates of the point where the graph crosses the y-axis
  b the coordinates of the minimum point of the graph.

5 Using your answers to question **2**, write down:
  a the coordinates of the point where the graph crosses the y-axis
  b the coordinates of the minimum point of the graph.

PS 6 Using your answers to question **3** to complete these questions.
  a Write down the coordinates of the minimum point of the graph.
  b Write the equation $x^2 + 4x - 6 = 0$ in the form $(x - a)^2 + b = 0$
  c What is the connection between minimum point and the values in the equation when it is written as $(x - a)^2 + b$
  d Without drawing the curve, predict the minimum point of the graph $y = x^2 + 6x - 5$

AU 7 Beryl draws a quadratic graph which has a minimum point at $(2, -6)$.
  She forgets to label it and later cannot remember what the quadratic function was.
  She knows it is of the form $y = x^2 + px + q$.
  Can you help her?

Problem-solving Activity

Drawing linear graphs

a Draw the line $y = 2x + 3$ on a grid like the one shown here.

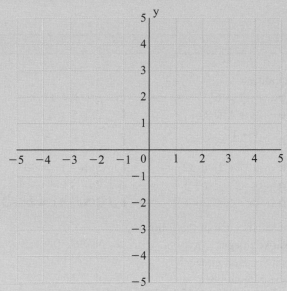

Draw the line $y = -\frac{1}{2}x + 3$ on the same grid.
What do you notice about the angle between the lines?

b Draw the lines $y = -4x + 1$ and $y = \frac{1}{4}x + 1$ on the grid.
What do you notice about the angle between the lines?

c Draw the lines $y = \frac{2}{3}x - 1$ and $y = -\frac{3}{2}x - 1$ on the grid.
What do you notice about the angle between the lines?

d Look at the equations of the lines in parts **a**, **b** and **c**.
What is the connection between the gradients in each part?

e Write down a rule for the gradients of perpendicular lines.

Algebra: Inequalities and regions

15.1 Solving inequalities

HOMEWORK 15A

1 Solve the following linear inequalities.

a $x + 3 < 8$	**b** $t - 2 > 6$	**c** $p + 3 \geqslant 11$
d $4x - 5 < 7$	**e** $3y + 4 \leqslant 22$	**f** $2t - 5 > 13$
g $\dfrac{x+3}{2} < 8$	**h** $\dfrac{y+4}{3} \leqslant 5$	**i** $\dfrac{t-2}{5} \geqslant 7$
j $2(x - 3) < 14$	**k** $4(3x + 2) \leqslant 32$	**l** $5(4t - 1) \geqslant 30$
m $3x + 1 \geqslant 2x - 5$	**n** $6t - 5 \leqslant 4t + 3$	**o** $2y - 11 \leqslant y - 5$
p $3x + 2 \geqslant x + 3$	**q** $4w - 5 \leqslant 2w + 2$	**r** $2(5x - 1) \leqslant 2x + 3$

2 Write down the values of x that satisfy each of the following.

a $x - 2 \leqslant 3$, where x is a positive integer.

b $x + 3 < 5$, where x is a positive, even integer.

c $2x - 14 < 38$, where x is a square number.

d $4x - 6 \leqslant 15$, where x is a positive, odd number.

e $2x + 3 < 25$, where x is a positive, prime number.

3 Frank had £6. He bought three cans of cola and lent his brother £3.
When he got home he put a 50p coin in his piggy bank.
What was the most the cans of cola could have cost?

4 Solve the following linear inequalities.

a $9 < 4x + 1 < 13$	**b** $2 < 3x - 1 < 11$	**c** $-3 < 4x + 5 \leqslant 21$
d $2 \leqslant 3x - 4 < 15$	**e** $10 \leqslant 2x + 3 < 18$	**f** $-5 \leqslant 4x - 7 \leqslant 8$
g $3 \leqslant 5x - 7 \leqslant 13$	**h** $8 \leqslant 2x + 3 < 19$	**i** $7 \leqslant 5x + 3 < 24$

AU 5 The perimeter of this rectangle is greater than 10 but less than 16.

$2x - 1$

x

What are the limits of the area?

PS 6 A teacher asks six students to stand at the front of the class and hold up the following cards.

$x > 0$	$x < 2$	$x \geqslant 3$	$x = 2$	$x = 3$	$x < 9$

She writes 'TRUE' on one side of the board and 'FALSE' on the other side.
She asks the other students to call out a number, and the students holding the cards have to stand on the 'TRUE' side if their card is true for the number, or on the 'FALSE' side if it isn't.

a A student calls out '2' and the students all go to the correct side.

    i Which cards are held by the students on the 'TRUE' side?

    ii Which cards are held by the students on the 'FALSE' side?

b Find a value that would satisfy this grouping:

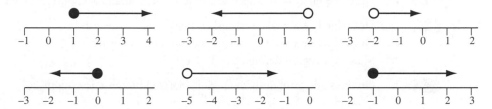

True

$x \geqslant 3$	$x < 9$	$x > 0$

False

$x < 2$	$x = 2$	$x = 3$

HOMEWORK 15B

1 Write down the inequality that is represented by each diagram below.

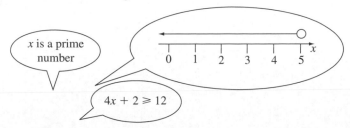

2 Draw diagrams to illustrate the following.

a $x \leqslant 2$ **b** $x > -3$ **c** $x \geqslant 0$ **d** $x < 4$

e $x \geqslant -3$ **f** $1 < x \leqslant 4$ **g** $-2 \leqslant x \leqslant 4$ **h** $-2 < x < 3$

3 Solve the following inequalities and illustrate their solutions on number lines.

a $x + 5 \geqslant 9$ **b** $x + 4 < 2$ **c** $x - 2 \leqslant 3$ **d** $x - 5 > -2$

e $4x + 3 \leqslant 9$ **f** $5x - 4 \geqslant 16$ **g** $2x - 1 > 13$ **h** $3x + 6 < 3$

i $3(2x + 1) < 15$ **j** $\dfrac{x + 1}{2} \leqslant 2$ **k** $\dfrac{x - 3}{3} > 7$ **l** $\dfrac{x + 6}{4} \geqslant 1$

FM 4 Mary went to the record shop with £20. She bought two CDs costing £x each and a DVD costing £9.50. When she got to the till, she found she didn't have enough money. Mary left the DVD and paid for the two CDs.

On the way home, she had enough money to buy a lipstick for £7.

a Explain why $2x + 9.5 > 20$ and solve the inequality.

b Explain why $2x + 7 \leqslant 20$ and solve the inequality.

c Show the solution to both of these inequalities on a number line.

d If the price of a CD is a whole number of pounds, how much does it cost?

AU 5 Copy the number lines below and draw two inequalities on them so that they have the integers {5, 6, 7, 8} in common.

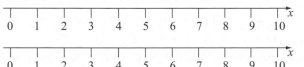

PS 6 What number is being described?

x is a prime number

$4x + 2 \geqslant 12$

7 Solve the following inequalities and illustrate their solutions on number lines.

a $\dfrac{5x+2}{2} > 3$ **b** $\dfrac{3x-4}{5} \leqslant 1$ **c** $\dfrac{4x+3}{2} \geqslant 11$ **d** $\dfrac{2x-5}{4} < 2$

e $\dfrac{8x+2}{3} \leqslant 2$ **f** $\dfrac{7x+9}{5} > -1$ **g** $\dfrac{x-2}{3} \geqslant -3$ **h** $\dfrac{5x-2}{4} \leqslant -1$

15.2 Graphical inequalities

HOMEWORK 15C

1 **a** Draw the line $y = 3$ (as a solid line).   **b** Shade the region defined by $y \geqslant 3$

2 **a** Draw the line $x = -1$ (as a dashed line).   **b** Shade the region defined by $x < -1$

3 **a** Draw the line $x = -1$ (as a dashed line).
    b Draw the line $x = 3$ (as a solid line) on the same grid.
    c Shade the region defined by $-1 < x \leqslant 3$

4 **a** On the same grid, draw the regions defined by these inequalities:
     i $-2 \leqslant x \leqslant 2$ **ii** $-1 < y \leqslant 3$
    b Are the following points in the region defined by both inequalities, above?
     i $(2, 2)$ **ii** $(-2, 2)$ **iii** $(-2, -1)$ **iv** $(-2, 3)$

5 **a** Draw the line $y = 2x + 1$ (as a solid line).
    b Shade the region defined by $y \leqslant 2x + 1$

6 **a** Draw the line $3x + 4y = 12$ (as a dashed line).
    b Shade the region defined by $3x + 4y > 12$

7 **a** Draw the line $y = 2x - 1$ (as a solid line).
    b Draw the line $x + y = 5$ (as a solid line) on the same diagram.
    c Shade the diagram so that the region defined by $y \geqslant 2x - 1$ is left unshaded.
    d Shade the diagram so that the region defined by $x + y \leqslant 5$ is left unshaded.
    e Show clearly with an R the region defined by both inequalities.

8 **a** Shade $x \leqslant 2$, $y \geqslant x - 2$ and $x + y \geqslant -2$ on the same grid.
    b Show clearly the region defined by all three inequalities by a letter R.

9 **a** On the same grid, draw the regions defined by the following inequalities. (Shade the diagram so that the overlapping region is left blank.)
     i $y < x + 2$ **ii** $y \geqslant 2x - 2$ **iii** $y \geqslant 0$
    b Are the following points in the region defined by all three inequalities?
     i $(0, 2)$ **ii** $(0, -2)$ **iii** $(2, 2)$ **iv** $(4, 4)$

10 **a** On a pair of axes, leave unshaded the region represented by the following inequalities: **i** $x \leqslant 2$ **ii** $y > 1$ **iii** $y \leqslant x + 1$
    b Write down the coordinates of all the points whose coordinates are integers and lie in the region that satisfies all the inequalities in part **a**.

AU **11** The graph below shows three points, (1, 1), (2, 1) and (2, 2).

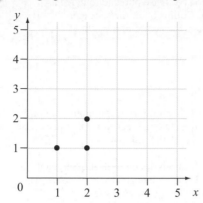

Write down three inequalities that between them surround these three grid intersection points and **no others**.

PS **12** If $x - y \leq 50$, which of the following may be true (M), must be false (F) or must be true (T)?

a $x > 20$ **b** $x + y > 60$ **c** $x - y = 10$ **d** $y \leq 0$

e $x + y > 50$ **f** $y > x - 50$ **g** $y = x$ **h** $x + y < 51$

Functional Maths Activity

League champions

Albion Rovers are at the top of a football league. They have played all their games and have 58 points. The only other team that could come top of the league are Albion United. They currently have 50 points, with four games still to play.

Teams are awarded three points for a win, one point for a draw and no points for a loss.

Alice is a keen United fan and she draws this graph to show the possible outcomes:

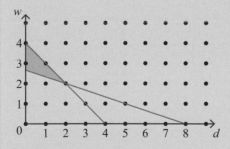

She also writes down:

$w + d \leq 4$

$3w + d \geq 8$

She says that the shaded region shows what United need to do to become league champions.

1. What do you think w and d stand for?
2. Can you explain why she has written down the two inequalities and why she has shaded the region on the graph?
3. Describe in words what United need to do to win the league.
4. How would the graph be different if United had five games still to play instead of four?

William Collins' dream of knowledge for all began with the publication of his first book in 1819. A self-educated mill worker, he not only enriched millions of lives, but also founded a flourishing publishing house. Today, staying true to this spirit, Collins books are packed with inspiration, innovation and practical expertise. They place you at the centre of a world of possibility and give you exactly what you need to explore it.

Collins. Freedom to teach.

Published by Collins
An imprint of HarperCollins*Publishers*
77–85 Fulham Palace Road
Hammersmith
London
W6 8JB

Browse the complete Collins catalogue at
www.collinseducation.com

© HarperCollins*Publishers* Limited 2010

10 9 8 7 6 5 4 3

ISBN-13 978-0-00-734027-9

Brian Speed, Keith Gordon, Keith Evans, Trevor Senior and Chris Pearce assert their moral rights to be identified as the authors of this work

All rights reserved. No part of this publication may be reproduced, stored in a retrieval system, or transmitted in any form or by any means, electronic, mechanical, photocopying, recording or otherwise, without the prior written permission of the Publisher or a licence permitting restricted copying in the United Kingdom issued by the Copyright Licensing Agency Ltd., 90 Tottenham Court Road, London W1T 4LP.

British Library Cataloguing in Publication Data
A Catalogue record for this publication is available from the British Library

Commissioned by Katie Sergeant
Project managed by Patricia Briggs
Edited by Brian Asbury
Answers checked by Steven Matchett and Joan Miller
Cover design by Angela English
Concept design by Nigel Jordan
Illustrations by Wearset Publishing Services
Typesetting by Wearset Publishing Services
Production by Leonie Kellman
Printed in China

Important information about the Student Book CD-ROM
The accompanying CD-ROM is for home use only. You cannot copy or save the files to your hard drive and it will work only when placed in the CD-ROM drive.